The Modern Technical Writer

C. Edward Collins

Prentice Hall Canada Career & Technology
Scarborough, Ontario

Canadian Cataloguing in Publication Data

Collins, C. Edward, 1933-
   The modern technical writer

ISBN 0-13-899444-7

1. Technical writing.   2. English language – Technical English.   I. Title.

T11.C653 1998   808'.0666   C97-931242-6

 © 1998 Prentice-Hall Canada Inc., Scarborough, Ontario
A Division of Simon & Schuster/A Viacom Company

Prentice-Hall, Inc., Upper Saddle River, New Jersey
Prentice-Hall International (UK) Limited, London
Prentice-Hall of Australia, Pty. Limited, Sydney
Prentice-Hall Hispanoamericana, S.A., Mexico City
Prentice-Hall of India Private Limited, New Delhi
Prentice-Hall of Japan, Inc., Tokyo
Simon & Schuster Asia Private Limited, Singapore
Editora Prentice-Hall do Brasil, Ltda., Rio de Janeiro

ISBN: 0-13-899444-7

Acquisitions Editor: David Stover
Editorial Assistant: Ivetka Vasil
Production Editor: Amber Wallace
Copy Editor: Gail Copeland
Production Coordinator: Jane Schell
Cover/Interior Design: Monica Kompter
Cover Image: Todd Davidson/The Image Bank
Page Layout: Phyllis Seto

1 2 3 4 5   RRD   02 01 00 99 98

Printed and bound in the United States

Visit the Prentice Hall Canada Web site! Send us your comments, browse our
catalogues, and more at **www.phcanada.com**. Or reach us through e-mail at
phabinfo_pubcanada@prenhall.com.

This book is dedicated to John C. McDonald:
talented writer and artist, true friend.

Contents

CHAPTER 5   The Technical Writer's Tool Kit   56

SECTION III — THE FORMS   133

CHAPTER 9   Short Reports   138

CHAPTER 10   Long Reports   173

CHAPTER 11   External Proposals   206

CHAPTER 12   Instructions   236

CHAPTER 13   Correspondence   253

CHAPTER 14   Oral Presentation   275

CHAPTER 15 Job Search Techniques   289

SECTION IV — ETHICS   305

CHAPTER 16   Ethics and the Technical Writer   306

Ethical Foundations   307

SECTION V — A SELF-DIRECTED COURSE IN ENGLISH GRAMMAR AND USAGE   331

PART A   Grammar Fundamentals   332

An Overview   332

The Parts of Speech   332

The Characteristics of Verbs   335

The Elements of the Sentence   338

PART B   An Introduction to English Usage   347

Preface

Rationale

Technical writing is a rapidly-changing and complex field, and, for the most part, authors and publishers of technical writing textbooks have yet to fully acknowledge the nature of the work that modern technical writers actually do. In fact, researchers in the field have even used the phrase "school-based fantasy about the workplace" in reference to some of the textbooks currently on the market.

For this reason, I have chosen to avoid the well-worn path travelled by other authors and to offer students a look at the real world of technical communication. In pursuing this end, I have acknowledged three important facts:

- Most technical documents are associated with getting a product or service from a perception of its need to the marketplace.
- Major technical documents are most often produced by a team of people.
- Although writing good technical documents is hard work, there is a best way to go about it.

The Modern Technical Writer is based on yet another important truth: technical communication is a *technical* subject. This fact does not imply that one need only learn a few formulas for writing technical documents. On the contrary, a proper analysis of technical communication leads to the opposite conclusion: technical writing is a complex subject. However, the analysis of the writing process contained in this book simplifies, insofar as simplification is possible, the technical communicator's job. A thorough and, incidentally, unique exposition of the writing process lies at the heart of *The Modern Technical Writer.*

Student Diversity

Not all students have a natural talent for using language, and language preparation levels vary greatly among students in today's classrooms. Nevertheless, I think that nearly all students who approach the subject of technical communication as it is presented in this book can become able technical communicators.

Students can learn the principles of technical communication in the same way they learn the principles of statics, DC circuits, or automotive mechanics. Nearly all technical communication students, even those who point out that "English was my worst subject in high school," can learn to write good technical documents by working through each step in the writing process, which is neither mysterious nor hard. Indeed, technical communication is presented herein as a subject that anyone with average verbal ability can master.

A Real-World Approach

The Modern Technical Writer is based on the assumption that the best way to prepare for real-world writing and speaking tasks is to practice the kinds of communication one will face on the job. A recent survey disclosed that 90 percent of all technical communicators work in the aerospace or electronics industries, approximately two-thirds of them in manufacturing.

In order to provide opportunities for students to develop and practice the skills they will need after graduation, the book includes the following special features:

Extensive Exercises Based on Real-World Situations

Most of the exercises simulate communication problems similar to those the graduate will encounter in the working world. Although realistic, *the exercises do not require technical knowledge*. Many of the exercises offer opportunities for students to respond as individuals and as members of writing teams.

A Typical Writing Specification and a Typical Style Guide

Students are introduced to the concept of writing formal documents that must reflect adherence to a specification. The typical specification answers questions such as the following: How do I refer to and place illustrations? What heading format and page layout should I use? Where should I place the front matter items such as the abstract and table of contents? Where should I put the page numbers?

Writing conventions, which vary from one firm to another, are introduced in the typical style guide, as are some standard styles for writing numbers, symbols, etc.

Chapters on the Electronic Office, Team Writing, and Ethics

The personal computer has changed the way technical writing gets done, and the chapter on the electronic office introduces technical students to desktop publishing and the role of World Wide Web in technical communication.

Since major technical documents are nearly always written by groups of people, a chapter is devoted to the principles that govern collaboration, the composition of technical communications groups, and the methods by which groups proceed from document conception to printed pages.

People responsible for writing and publishing technical documents have an ethical obligation to their employers, their readers, and to the community at large. Simulated real-life ethical problems are presented in the exercises concluding Chapter 16. These exercises are based on cases submitted by members of The Society for Technical Communication. All of the ethics cases appeared in *INTERCOM*, the Society's monthly magazine.

A Self-Directed Course in English Grammar and Usage

A Self-Directed Course in Grammar and English Usage has been included for students who need to brush up on these skills. Also, Chapter 6, The Writing Process, includes grammatical analysis as one of the steps in the rewriting stage of document preparation, along with both rhetorical and logical analyses.

Acknowledgments

I am deeply indebted to the many people who helped in this book's publication. First of all, I wish to thank the staff of Prentice Hall Canada: Acquisitions Editor David Stover for his energy, enthusiasm, and faith in the project; Ivetka Vasil for her selection of such excellent reviewers; Amber Wallace for all her outstanding work throughout the production phase; Gail Copeland for her diligent copyediting of the manuscript; and Nick Gamble for his keen-eyed proofreading of the final pages.

My sincere thanks go The Society for Technical Communication's Maurice Martin and Anita Dosik, who secured permissions for some of the materials appearing in the Chapter 16. That material includes the ethics quiz written by Sam Dragga, ethics cases by John M. Gear and John G. Bryan, and case responses by David N. Parker, Jack Greenfield, Andrea Holloway, Paul S. McKelvey, Neil Lineberger, Deborah Snavely, Robert Lessman, and Jon Russell.

I would like to offer my very special thanks to John T. Cavanaugh, long-time friend and one of the most talented technical writers I've ever known, for writing the Appendix on HTML.

Thanks also to the following reviewers, whose experience and pedagogical knowledge helped immeasurably: John Anderson, St. Lawrence College; Nina Pyne, Sault College; Hugh Reed, Northern Alberta Institute of Technology; and George Tripp, Fanshawe College.

C.E.C.
1997

To the Student

If the Toronto Blue Jays were to practice with plastic balls and bats during spring training each year, they would be poorly prepared for the opening of the regular season, and their hopes to play in the World Series would likely be nothing more than an idle dream. Fortunately, the Jays play hardball all the time and start the season well prepared. This book does not present technical writing as a whiffle ball game either. A college student's preparation to think, organize, write, and speak well in the workplace has direct parallels to an athlete's practice for competition. Therefore, you will find that this book does not "talk down" to you. The vocabulary level is high, and you should expect to learn new words by looking them up as the need arises. Likewise, the exercises presented herein have nothing to do with adolescent self-discovery. Instead, they direct you to write real technical documents based on real work situations.

Graduates who pass through college without learning to think and write well are shocked awake as soon as they enter the world of international commerce. Often, they find that their awakening has come too late. On the other hand, students who accept the challenges of the exercises and apply the principles of the writing process presented in this book will see their thinking, writing, and speaking skills increase. They will gain a number of truly valuable skills and will become able to hold their own in the real world.

So, should you feel intimidated? The answer is no. This book presents technical communication as a technical subject, not as a study of loosely associated ideas or vague guidelines. If you work as hard as one must to learn any other technical subject, you will acquire the skills quickly.

About the Author

A Senior Member of the Society for Technical Communication and a Life Member of the Alberta Society of Engineering Technologists, C. Edward (Ed) Collins holds a degree in philosophy and has done graduate work in both epistemology and ethics. His technical writing career began in 1958 at Douglas Aircraft in El Segundo, California, where he wrote manuals to support carrier-based jet fighters. He continued in the aerospace industry for 11 years, working at Douglas Aircraft, the Westinghouse Astronuclear Laboratory, and as a technical writing consultant to several other aerospace firms. He has written and edited literally hundreds of technical documents, including the design verification test procedures for vital components of the Apollo Lunar Module and a six-volume study of the feasibility of a scientific observatory on the lunar surface. In addition to professional papers, he has written articles for a wide variety of magazines, including *The Wanderer, Flying, Kitplanes, DirecTalk,* and *Challenge.*

Mr. Collins also taught technical writing at the Northern Alberta Institute of Technology for many years and has authored or co-authored six other textbooks: *Plain English: A Guide to Standard Usage and Clear Writing, Plain English: Special Edition for the Canadian Department of Defence, The Basics of Technical Writing and Speaking, English at Work: Communicating in Business and Industry, Technical Communication at Work,* and *Modern Medical Language.* He continues to teach and also works as an independent technical communications consultant.

The
Foundations

CHAPTER 1

The Scope and Theory of Technical Communication

OBJECTIVES: In this chapter, you will learn that technical communication fulfills a variety of purposes, that written technical communication is called technical writing, and that although technical writing shares many of the characteristics of other kinds of writing, it has certain marks that set it apart. Along with your discovery of the three major divisions of technical writing, you will learn about general communication theory and will discover why and where technical writing is practiced within a company organization.

What Is Technical Writing?

A Simple Definition

Formal technical writing comprises three main categories: proposals, reports, and instructions. Its purpose is connected directly to the conception, design, manufacture, maintenance, or sales of products and services. Beyond these formal document categories, many other related activities occupy the modern technical writer. For example, technical writers are often called upon to write technical material for speeches and, sometimes, to deliver the speeches as well. Others write film scripts, advertising copy, or sales brochures. Most of these activities fall into the same three categories (proposals, reports, and instructions) and are related to a variety of business functions.

Contrary to what many non-technical writers believe, the reporting function makes up only about a third of technical writing. Approximately another third is taken up by proposals, in which the language of persuasion and the subtleties of tone are especially important. The final third is instructional, and this kind of writing has its own peculiar requirements.

The modern technical writer needs to know about all the categories of technical writing. This book describes the categories in detail. The writing process described in Chapter 6 may be applied to all the categories — and by everyone who has to write technical information.

The Relationship of Technical Writing to Other Kinds of Writing

Many kinds of writing stand apart from the purely literary forms such as novels and poetry. Conveyors of these non-literary kinds of writing include newspapers, magazines, advertising flyers, cereal boxes, textbooks, and many others. Among the others is the technical document, and the writing contained within such a document is called technical writing. Just how does this kind of writing differ from all the other kinds? How is it the same?

The answers to these two questions, very important for anyone embarking on a study of the subject of technical writing, are not hard to find. A moment of thought tells us that the purpose of a technical document is to inform, whereas the purpose of a novel is to entertain (in the broadest sense of that term). But that statement doesn't take us very far, since we will remember almost at once that the purpose of newspapers and some magazines is also to inform.

However, the inference that newspapers and magazines are technical writing is unjustified because readers of newspapers or magazines usually would not use the information in the same way as would readers of a technical document. The technical document has a narrower purpose than that of newspapers and most magazines, and readers are more likely to do something with the information in the technical document. Even so, we can safely state that the shared purpose of informing makes technical documents closer to newspapers and magazines than they are to novels or poetry. In fact, some magazines, such as *Kitplanes, McCall's Quilting, Popular Electronics*, and others, are first cousins of formal technical documents.

How Can We Recognize Technical Writing?

SINGLE MEANING

Continuing our consideration of the purpose of technical writing, we will see that single meaning is also a characteristic of the technical document. Oddly enough, this need to avoid ambiguity bears at least one similarity to good poetry. Both the poet and the technical writer try to use words as sparingly as possible, although the poet is trying to say something that cannot be conveyed in ordinary prose and the technical writer might be trying to give foolproof instructions for replacing a faulty hydraulic valve. A big difference between a technical document and a poem is that the meaning in a technical document does not lie beyond the words, as the meaning of a poem might well do. The words and structures a technical writer chooses will lend themselves to meaning in a direct and straightforward way. The well-written technical document is not only unambiguous, it is also in complete accord with the physical facts. It is *accurate*.

Let's look at an example, taken from a maintenance manual, of a procedural step that could easily confuse readers:

Step 12:   X-ray inspect the empennage brackets which are different from the structural brackets attached at station y-427.

Here we have a simple, but confusing, instruction. The confusion comes about because the writer was careless. A journalist or an advertising copywriter might omit the commas that separate a non-restrictive adjective clause (introduced by the word *which* in the example) from the rest of the sentence, but the technical writer who does the same thing courts disaster. Applying a comma rule, the writer would have written the step as follows:

> Step 12: X-ray inspect the empennage brackets, which are different from the structural brackets, attached at station y-427.

Good punctuation solves the problem, but if we look at the instruction from the technician's point of view, we will see that Step 12 will benefit from further improvement. Some readers might wonder whether all the empennage brackets *and* all the structural brackets are located at Station y-427. Furthermore, the step must include an illustration so that the reader may differentiate between the two kinds of brackets. The following version is accurate (and foolproof) because it anticipates and answers all the technician's questions:

> Step 12: X-ray inspect the empennage brackets, which are attached only at station y-427. The structural brackets, which are different from the empennage brackets (see Figure 17-18), are located at station y-427 and throughout the fuselage assembly. Do not inspect the structural brackets.

The first version of the step tells the technician to differentiate between the structural brackets and the empennage brackets, but it neither gives the location of the brackets to be inspected nor refers the reader to an illustration that identifies the brackets to be inspected. For these reasons, the reader might attempt to interpret the procedure and make a mistake that could create a hazard to life and equipment.

The second version of the step would, more than likely, lead to wasted time rather than to disaster. The technician would probably have to identify the empennage brackets by referring to a parts manual.

Many errors similar to those just illustrated have produced disastrous consequences. When the first Gemini spacecraft was launched, it fell into the sea instead of achieving orbit. The mission failed because the technical writer had written an ambiguous step in the preflight procedure. Fortunately, since there was no crew aboard the spacecraft, no lives were lost. In another instance, however, many people lost their lives because the author of a service change bulletin gave incomplete instructions for replacing an empennage part in an airliner. The mechanic who performed the service change guessed wrong, and the airliner plunged to the ground, killing all aboard.

Here is another example from a real technical manual. A technical writer wanted to write a procedural step directing the user to insert a pin into an assembly. Since no tools were required for this step, the writer wrote: "Insert pin by finger." The word choices here were perhaps not as good as they should have been, but no damage would have been done had not the manuscript typist left out a space, thereby converting the step to "Insert pinby finger." Unfortunately, this typing error made it through the production department's proofreading process. The compositor, puzzling over the word "pinby," finally concluded that the step should read: "Insert pinky finger." The step appeared that way in the printed manual, causing embarrassment to the company and placing in jeopardy the little fingers of compliant and unthinking technicians.

ECONOMY AND CLARITY

We now have answers to the questions posed earlier. Technical writing informs readers about technical subjects. Its language should be unambiguous, and its structure should be verbally economical and clear. It resembles newspapers and magazines in that its function is to inform, but it differs from them in that its purpose is to enable a reader to perform a task or to make a business decision.

The qualities of economy and clarity are extremely important. However, they cannot be achieved directly. Writing that attempts to be economical often deteriorates to terseness. And efforts to write clearly can sometimes produce documents in which the author appears to be "talking down" to the reader. The careful writer avoids these pitfalls by means of attitude and approach. If these are correct, the desirable qualities of economy and clarity come about naturally as a by-product.

What Departments Generate Technical Writing?

The technical writing function is found throughout the structure of any business firm. Most large companies have a technical publications department through which all technical documents are channeled. In some small companies, each person responsible for originating technical documents must be able to do the entire job of writing, editing, page layout, and final publication. In very small companies, the same person may manufacture and deliver the products — and maybe even sweep the floor at the end of the day. However, no matter what the company's size, all the information needed for a product to reach the marketplace must be generated, and someone must generate it.

The functions needed to produce this information are best understood in the context of a complete organizational structure.

Let's look now at the organization chart of a typical firm. (See Figure 1-1.)

FIGURE 1-1 | Typical Organization Chart

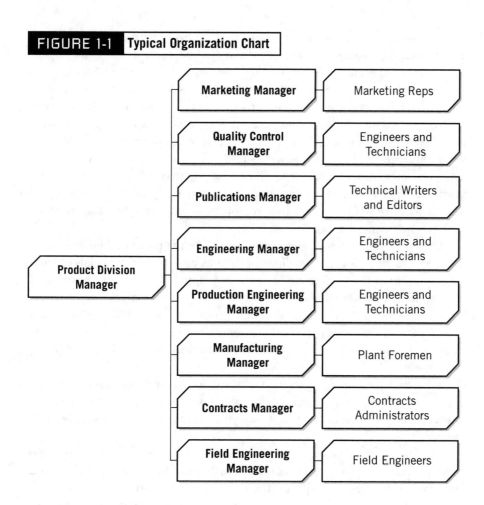

Marketing Manager	Marketing Reps
Quality Control Manager	Engineers and Technicians
Publications Manager	Technical Writers and Editors
Engineering Manager	Engineers and Technicians
Production Engineering Manager	Engineers and Technicians
Manufacturing Manager	Plant Foremen
Contracts Manager	Contracts Administrators
Field Engineering Manager	Field Engineers

Product Division Manager

Marketing

Marketing personnel determine how best to sell their company's products and endeavor to learn what new products are needed in the marketplace. In carrying out the latter function, a marketing department acts as a clearinghouse for information passing between the executives and engineers in its own company and their counterparts in a potential customer's organization. The technical writing that goes on in this area comprises *product data sheets, capabilities statements, proposals,* and many different kinds of *memorandum reports.* Note that these documents are not always written by marketing representatives. They get help from other departments, especially the publications department. In fact, as we will see, the technical writing function is one that requires constant liaison and collaboration (see Chapter 3). Let's look at each document type in turn.

Product data sheets are a very important means of getting information to a potential customer. They are of primary interest to the technical person who wants quick information for easy comparison of your product to your competitor's product. Technically trained people, such as engineers and technicians, do not want a lot of "gloss"; they want facts. Therefore, the secret to writing good data sheets is to ask yourself what you as a knowledgeable buyer would want to know. Product brochures, a type of advertisement, are most often put together by someone with a knowledge of advertising. As a technician, engineer, or technical writer, however, you may be asked to contribute copy that might be used in a brochure. Sometimes, report excerpts, too, find their way into brochures, another reason that enlightened managers insist on clear and concise reports.

Capabilities statements are reports on the product history and the physical facilities of a firm. Some capabilities statements also contain financial data and information on the backgrounds of key personnel. The primary purpose of the capabilities statement is to convince a potential customer (such as a government agency or a major firm) to place your firm on its bidder's list. Then when the procurement officers offer a contract for goods or services, or when they send out requests for proposals, they will consider your firm as a possible supplier.

Proposals are quite often major technical documents, and as such they require the combined efforts of many people. Some middle-size companies produce as many as 50 major proposals a year, aiming for a success rate of 10 percent.

Memorandum reports are the mainstay of interoffice communication. Of course, every memorandum is not necessarily a report, but most of those you will write generally will fall into the report category. These documents are important to you apart from their subject matter. Your personality and organizational abilities are visible in your writing, and if you are interested in promotions, you should also be interested in learning to write good memos.

Engineering

A "line item" in a contract refers to the goods to be delivered under the terms of the contract. Contracts routinely contain line items that cite the delivery dates for technical documents. For example, if your company were to contract to carry out a 12-month development program for a new piece of equipment, the line items in the contract might read as follows:

1. Interim reports to be submitted March 31, June 30, September 30.

2. Calibration and test procedures for preproduction models to be submitted by October 31.

3. Ten preproduction models to be submitted for test by October 31.
4. Final Report to be submitted by December 31.
5. First batch of production models to be submitted by January 15.
6. Handbook of Operating and Maintenance Instructions to be submitted by January 15.

Four of the six items in the foregoing example are for documentation, indicating clearly that the technical writing task would make up a major part of the work to be done. It is also clear that technical writing is not a peripheral activity, to be worried about after the "important" items are done. Too many newly established companies have found out too late that such is not the case.

Many companies require their technical people to keep a daily logbook recording progress on the various tasks they perform. This daily log can have a personal character, in that most people develop their own shortcuts and ways of referring to their work. However, clear and concise logbooks are very important when report time comes around.

Internal reports are often required as a record of work done, especially for projects funded "in-house." These reports are often, but not always, written in a memorandum format.

Quality Control

Test procedures of all kinds are essential for the efficient evaluation not only of the products that a company builds, but also of the components and subassemblies it buys. An electronics firm, for example, may buy literally millions of electronics components for use in the products it builds. A way of estimating the reliability of these components must be in place. Therefore, companies must establish sampling procedures and rigorous testing programs to assure that faulty components or subassemblies will not compromise the quality and reliability of the equipment they build. Such procedures and programs represent a major writing task.

Publications

The publications department sometimes acts as a service organization to other departments, providing writing, editing, and production help to the departments that originate information. The publications department in some firms, however, assumes the responsibility for all documentation. In either case, close cooperation must be maintained between the publications department and the other line departments if the publications function is to be carried out successfully.

Manufacturing

Manufacturing processes must be delineated clearly, so that those in charge of setting up production lines can know exactly how to go about their jobs. In addition, some companies even insist on daily or weekly production reports.

Field Engineering

People working in the field also write reports as a regular part of their jobs. These reports are useful not just for supervisors who have to keep track of what their field personnel are doing, but also for the engineering and quality assurance departments. Patterns of equipment failures can be discerned from field reports, and subsequent service changes can be effected to make products more reliable.

Summary

Formal technical communication encompasses the three main forms of technical writing: proposals, reports, and instructions. Technical documents are required at every stage in the life of a product or service — from perception of need to training and maintenance in the field. Most engineers, technologists, and technicians need to know about the kinds of documents their companies generate. And since technical writing is an integral part of every firm's operation, many technicians, technologists, and engineers must be able to contribute to technical documents.

A Theory of Communication

Communication, the most common and at the same time the most difficult of all human activities, is blamed for everything from crime to family breakups and world wars. What makes it difficult? Are communication techniques complex, like the principles of nuclear science?

Contrary to popular belief, communication *techniques* are usually not the cause of communication failure. Most difficulties come about when communicators misunderstand the relationship between thought and language or when the ongoing conflict between human will and logic tips in favor of human will, thereby opening the door to unethical practices. Let's look first at the relationship between thought and language.

Thought and Language

When we think about physical objects or events, we might see pictures, either still or moving, or we might see or hear words. Many primitive cultures developed systems of writing with pictures and discovered a grammar enabling them to put pictures together in such a way that fairly complex ideas could be recorded. It is interesting to note that grammar is not only a consistent feature of all languages, but that its basic structure is the same for all languages. Although usage rules vary from culture to culture, grammar is rooted in a fundamental understanding of the world, an understanding common to all cultures. It is also interesting to note that the languages of primitive cultures are not the least bit primitive.

A further development of writing systems came about with alphabets, which allow sounds to be converted into visual symbols. A by-product of this advanced system of writing was that thinkers could organize their thoughts more rapidly and see the relationships between them with greater clarity than they could before. Almost at once, thinkers began to work in abstract terms and to "see" reality in those same terms. This process of abstraction also made it easier for people to think, talk, and write about non-physical realities, such as love, hate, logic, law, etc. Figure 1-2 shows the chronological development of language.

FIGURE 1-2   **The Chronological Development of Language**

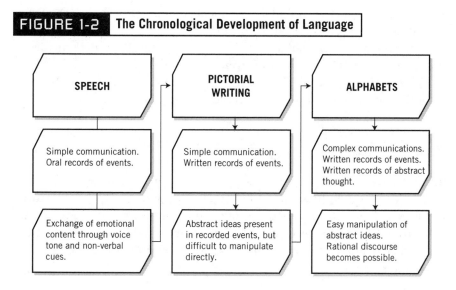

By using words, Plato and Aristotle were able to discover important principles of human action and interaction. Descartes discovered philosophical and mathematical principles in the same way, and Newton and Einstein conceived their theories of the physical universe by thinking in words and mathematical symbols.

Since the very act of thinking depends on language, one can conclude that discoveries come about because of language and that most such endeavors exist quite apart from the discoverer's wish to communicate with others. In fact, without this ability to cast thoughts in abstract symbols, the human race could never have progressed to current levels of civilization and technology.

LANGUAGE AND THOUGHT GO HAND IN HAND

So what's the problem? How does the foregoing discussion show that communication is difficult? The problem occurs when we fail to realize that we cannot separate the communication and thought processes. When we try, we come up with a variety of methods for fixing up faulty communication, but even when the methods work — and they most often don't work — we gain nothing in communicating muddled thoughts.

Since language and complex thought are inseparable, we need to look at words and language structures as they affect our thinking, and as they affect our communication with others. The processes of thought and communication are so similar that we can often clarify our thinking by writing our thoughts on paper and by reading them to see our own errors in logic.

In a famous essay on the relationship between politics and language, George Orwell pointed out that if we have foolish thoughts, we make foolish statements, and then these foolish statements reinforce our foolish thoughts. Orwell says that this cycle can be likened to the man who begins drinking abusively because he is a failure. And then his drinking affects his activities to such an extent that he becomes an even bigger failure.

Orwell believed that speaking and writing without thinking can become a habit that eventually destroys one's power to think.

WORDS HAVE POWER

If words have enabled people to become civilized and to advance technologically, we can conclude that they are both powerful and important. Unfortunately, we sometimes forget how important words are. Semanticists (students of the meanings of words) often remind us that words are not the things they refer to. They are right of course, but we must never forget that a word is a *thing* too.

We should never think of words as unimportant. Instead, we must understand that although many of the symbols used to signify reality are arbitrarily chosen, words have a reality and a power of their own. This power derives from several sources, including the general agreement concerning the meanings of words.

As an illustration of how this general agreement lends power to words, let us consider the word *fire*. We might have had another name for the phenomenon referred to by this word; we might have called it *parsley*, and we might have called the herb we know as parsley *fire*. If the meanings of those two words *were* reversed, walking into a crowded theater and shouting "parsley!" would create panic. But beyond the curious interest one might have in knowing how various phenomena and objects acquired their names, one might ask: "So what?"

And the question is a good one. A lot of people believe that only poets and other language stylists should care about such matters. On the other hand, suppose someone yells *fire* in a crowded theater and people get trampled to death as the crowd rushes for the exits. Picture the courtroom scene during the subsequent trial for malicious mischief, when the defendant takes the witness stand.

Defendant's lawyer:	"Tell us exactly what happened on the evening in question."
Defendant:	"I was upset because the garnish had been left off my dinner in the restaurant across the street from the theater. I was still angry about this when I walked into the theater, and so I shouted *fire*."
Defendant's lawyer:	"What you really meant to shout, then, was *parsley*. You certainly intended no mischief, did you?"
Defendant:	"No, I had no intention of causing such a fuss. Fire or parsley. What's the difference?"

The judge's only possible finding is, "guilty as charged."

The point is, of course, that we must respect the power of words and treat them as the objects of reality they are, even though their reason for being is to refer to other phenomena and other objects of reality.

THOUGHT, SPEECH, AND WRITING

The most important lesson to be learned from the foregoing discussion is that our writing and speaking habits are important to us even when we aren't trying to communicate with others. Semi-coherent speech and poorly organized written work do not merely reflect a sloppy mind; as George Orwell observed, they help create it.

The good news is that learning about technical communication and practicing its principles will provide benefits far beyond the boundaries of the classroom. You will find that your mental acuity will rise in proportion to the level of language mastery you achieve, opening up horizons you had perhaps not even dreamed of.

The bad news is that learning these principles and techniques requires time and commitment. But although there are no shortcuts, there is a direct route. That direct route is by way of systematic study and consistent application of what you learn. Have you heard the old joke about the tourist lost in New York who asks a person on the street how to get to Carnegie Hall? The punch line is: "Practice." And practice is the key to success in improving thought and communication habits, as well.

But how do you practice? Do you just sit down at your word processor and bash out something? When you have to give a technical speech, do you just stand up and muddle through as best you can, hoping to do better each successive time? Just as learning the rules of the road is a prerequisite in driver training, so too is learning a few rules of technical communication practice a prerequisite to successful communication.

The Value of a Routine

Systematic study must precede practice. And consistent application of the lessons learned in a systematic approach is the key that will open the door to good assignments and speeches. Each time you follow a specific routine, your work will improve. That routine is not a prescription for communication; it is a process that one goes through to bring out the best work one is capable of. Just as all professional golfers use a routine setup before striking the golf ball, good writers follow a process to release their inherent talents. You will learn about the writing process in Chapter 6.

A Communication Model

Communication among people can be likened to electronic communication. There is a transmitter, a medium containing the message, and a receiver. There can also be interference and equipment breakdowns. Figure 1-3 is a block diagram of communication flow.

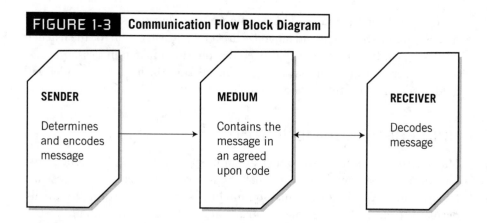

FIGURE 1-3 Communication Flow Block Diagram

SENDER	**MEDIUM**	**RECEIVER**
Determines and encodes message	Contains the message in an agreed upon code	Decodes message

The Sender

Apart from an intention to communicate, the sender must have an adequate measure of logic and language skills. Otherwise, communication will fail for technical reasons.

LOGIC SKILLS

The sender must use logic in putting the parts of the message into an understandable structure. Even if listeners and readers want to do so, they cannot assimilate details that do not fit into some recognizable pattern. And unless the message is clear and presented logically, even an interested listener will often tune out. Likewise, most readers will not spend extra time and effort to discover what illogically constructed writing has to say. Written messages must be carefully thought out and presented in a coherent structure. Chapter 5 contains important information about structural patterns and logical presentation.

LANGUAGE SKILLS

It is obvious that encoding a message requires a knowledge of the code to be used. Therefore, a thorough knowledge of Standard English, the code agreed upon by English speakers around the world, is necessary for anyone wishing to communicate in that language. When you make a grammatical (that is, structural) error in a computer program, the program will not run. But when you make a similar error in trying to communicate with another person, that person may very well attempt to compensate for the error and interpret the message. Such attempts often cause difficulty, and in technical writing, they can produce disastrous consequences.

Even if listeners and readers can figure out what the speaker or writer means, grammatical errors in language are a major distraction for the listener or reader. Garbled messages that need deciphering make unnecessary work for the receiver, work that may not get done because the listener or reader has neither the time nor the inclination to complete the job of communication.

NONVERBAL COMMUNICATION

Major distractions can also come from nonverbal sources. You might be reluctant to deposit your money in a bank that has a manager who dresses and behaves like a circus clown. And you might hesitate before voting for an aldermanic candidate who shows up drunk at a political rally. Such nonverbal *faux pas* are rare and obvious, but less drastic nonverbal cues to a speaker's or writer's credibility are always present. Therefore, appearance and demeanor are of great importance in oral communication settings.

Written communication also reflects the sender's demeanor and attitude toward the receiver. A document with coffee stains, bad handwriting, or other signs of inattention to graphic appeal says that the sender doesn't care much about the receiver.

A second benefit accrues from an attention to appearance. You will take yourself more seriously if you project an image that lends seriousness to your message. In the early days of radio, announcers dressed formally to read the news, even though they knew perfectly well that they would not be seen by their listeners. The BBC, the CBC, and other networks insisted that their announcers dress in a way that the network executives wanted them to *sound.* In the early days of radio, studio musicians also dressed in fashions consistent with the kind of music they were playing.

OBSTACLES TO COMMUNICATION ASSOCIATED WITH THE SENDER

From the foregoing discussion, we can identify four possible obstacles to communication, all of which can be eliminated by the sender. In addition to these, a number of secondary obstacles, which the sender may not be able to eliminate or control, can affect the outcome of communication attempts.

Primary Obstacles

1. Lack of intention to communicate
2. Lack of logic skills
3. Lack of language skills
4. Lack of awareness of nonverbal communication

Secondary Obstacles

1. Fatigue
2. Unavailability of time
3. Unavailability of an adequate work space
4. Unavailability of the proper tools

Most employees who have more work than can be reasonably expected of them do poor work, whether the work is laying concrete or writing reports. Likewise, employees who are forced to write business documents on poorly maintained or out-of-date equipment, utilizing their spare time and writing in poorly lit, crowded offices, will not do the best job they are capable of doing. However, an employee faced with such a situation may have little to say about it.

Often, only an enlightened management can remove secondary obstacles. Too often, management's response to poor business communications within its own firm is to send the employees to evening classes or to quick-fix report writing seminars. But all of the report writing courses in the world will not make up for the effects of a poor working environment.

The Medium

One of the first steps in the communication process is to decide which of the three categories (proposal, report, or instruction) your communication falls into and which medium should be used to convey it. Should the message be conveyed in a memorandum, a long document, an oral presentation, a telephone conversation, or some other communication mode? Selection of the wrong medium will often result in a failure to communicate. If your boss telephones to ask you to look up an address, a return telephone call to relay the information would be a good choice, and a memorandum would definitely be out of place. On the other hand, a telephone call reporting to your boss how the project you have been working on for the past six months relates to future business decisions would likewise be inappropriate unless it was immediately followed by an internal report (Chapter 9).

The Code

Standard English is the language of business throughout all of the English-speaking world. The varieties of English lying outside Standard English are many. For the linguist/sociologist studying the languages of regions and subcultures, all of these varieties may be of equal interest. However, when sociologists write reports on their findings, they will (if they are to enjoy credibility among educated readers) write them in Standard English. What is Standard English?

The answer to this question was much more easily given before 1961. That was the year that the third edition of *Webster's New International Dictionary* was published. The writers of this edition, adopting the position advocated by some modern linguists, did not think it democratic to distinguish between standard and non-standard usage. In their thinking, any pretense to "correctness" in language was grossly unfair to those who had not had the benefit of a formal education.

People who hold this view maintain that you speak correctly if you speak English. Since the language belongs to all of us, I have no right to correct you on matters of usage, and you have no right to correct me. (Presumably, we *both* enjoy the right to misunderstand each other.)

While the writers of this new dictionary may have had their hearts in the right place, they performed a great disservice to the users of English when they decided to base the new edition on this premise. First, although language does change constantly, most would-be changes fade away before they have a chance to become part of the language. Not many people would know your meaning if you said "23 skidoo" or "hubba hubba," both of which were widely used popular phrases, the first one in the 1920s and the second one in the 1940s. Second, communication cannot take place without agreement on the meanings of words and the rules of usage. And third, the people carrying on the business of the world must try to maintain standards whether or not philosophers, sociologists, or lexicographers (dictionary makers) sanction them.

Two results are clearly identifiable from such "modern" language theories. The social discrimination the reformers would have had us avoid has been increased by the disparities of usage among the various strata of society. And those who would prefer to pursue life goals outside their particular subcultures are confused by two messages. The message from the ivory towers of academe is that you must talk the way you have always talked; the message from the commercial world is that you must learn to speak and write Standard English if you want to live and work in the modern world.

The answer, then, to the question, "What is Standard English?", is that Standard English is English that follows rules of grammar, spelling, and syntax, and applies these rules consistently (whereas non-standard English picks and chooses which, if any, rules to follow, applying them inconsistently). Questions of usage can still be answered by *Webster's New International Dictionary, Second Edition*, by Fowler's *A Dictionary of Modern English Usage,* by Follett's *Modern American Usage,* and by a healthy regard for your own common sense.

The Receiver

The receiver needs the same skills as the sender: an intention to receive the message, a knowledge of logic, and a thorough understanding of the code, which in the English-speaking world is Standard English.

Note in Figure 1-3 that the arrow from the receiver points back toward the medium. The receiver must diligently attend to the medium, whether oral or written, because there is nothing automatic about the communication process. Successful communication always requires a joint effort by both the sender and the receiver.

Sometimes, the receiver can impose a structure on an unstructured message. Although this job does not properly belong at the receiving end of the communication process, a skilled listener or reader can often come up with a structure to help in understanding the message.

The following obstacles are most often responsible for communication failure at the receiving end:

1. Unwillingness to consider the message
2. Lack of logic skills
3. Lack of reading or listening skills
4. Lack of time or adequate work space

An unwillingness to consider the message can stem from several sources. You may be dissuaded by the sender's lack of language or logic skills, or you may be unmotivated because of preconceived notions about the sender or the message. If the communication fails because you refuse to wade through a maze of contorted reasoning and inflated, ungrammatical writing, you are hardly to blame.

On the other hand, if the communication fails because the speaker's accent reminds you of someone you don't like or because the writer's style is not exactly like yours, you must bear the responsibility for the failure. These obstacles can be easily overcome with a little self-discipline and practice.

Of course, the distractions of the crowded or otherwise unsuitable work space and the pressures of time can become obstacles at the receiving end of the communication process, just as they can at the sending end. But again, employees often can do little to remove these obstacles.

Summary

Nearly everyone can master basic communication techniques. However, communication often fails for reasons other than the poor technique exercised by the parties trying to communicate. Technical communicators who wish to communicate well must also learn about the nature of the communication process and must become aware of their responsibilities as technical communicators. The communication process also rests on the intentions of the parties to impart and receive information, a mutual agreement on the code to be used, and an acknowledgment of the fundamental principles of language structure.

Exercises

1-1. The Scope of Technical Writing

Write a brief paragraph to answer each of the following questions:

1. What are the three main categories of technical writing?

2. How does technical writing differ from other kinds of writing?

3. What are three important characteristics of technical writing?

4. Where in a company organization is technical writing generated? Fully explain your answer.

1-2. The Theory of Communication

1. Briefly discuss the reasons for breakdowns in communication.

2. Briefly discuss the relationship between thought and language.

3. Write a one-page paper discussing the responsibilities of the sender and receiver of a communication, along with the obstacles each may face.

CHAPTER 2

The Electronic Office

OBJECTIVE : In this chapter, you will become familiar with the tools available to the modern technical writer. The chapter is divided into three parts: communication networks, desktop publishing (techniques and tools), and the World Wide Web.

Communication Networks

The Computer Work Station

When computers were first introduced, they were used primarily for accounting and inventory control. A scant three decades later, the modern business firm now uses computers in nearly every part of its operation. Engineers use sophisticated programs to design and even to test new hardware. Many manufacturing operations are now controlled by computer-driven robots. And computer networks using either a LAN (local area network) or Modems (modulator-demodulator) connect employees both within and between departments and divisions of their companies. And the Internet, of course, provides connections to other individuals and institutions worldwide. Modern technical writers have, in a literal sense, a world of information at their fingertips.

Technical writers who have personal computers or computer terminals in their offices have a means of instant communication with fellow workers, as well as direct access to information from data bases that may be within the firm's own information bank or in some far-off facility that specializes in supplying research data.

The modern work station also facilitates other business tasks. For example, if the man seated at the work station shown in Figure 2-1 wishes to tele-

FIGURE 2-1	A Typical Work Station with Other Work Stations Shown in the Background

Source: Photo courtesy of IBM Canada Ltd.

phone someone, he doesn't even have to know the telephone number. He may merely open the telephone icon and, finding the name of the person he wishes to call, move it to the icon of the telephone. The computer then automatically dials the number of that person.

Electronic Mail

Written communication between members of a business firm has traditionally been in the memorandum format (which is discussed in detail in Chapter 9). Workers in modern offices can send and receive memorandums instantaneously without even getting up from their work stations. This exchange often takes place through a LAN of personal computers or computer terminals.

Some firms even handle external mail in this way by establishing an Internet link that can route electronic mail through telephone lines connected to central clearing computers located around the world. Computers are linked to these central clearing computers through Modems.

Fax Machines

A fax machine transmits printed copies of documents over telephone links. Fax (short for facsimile) machines have been around for many decades, but recent developments in fax technology have made them much faster and better. Fax machines can transmit documents, including drawings, photographs, and diagrams, to anywhere in the world in a matter of minutes.

Modern fax machines, which are available from many different manufacturers, combine other functions as well. The Xerox 3006 Plain Paper Fax/Printer, for example, is a fax machine, a photocopier, and a printer in one compact unit. With a PC Connectivity Kit option, it also becomes a scanner.

As a plain paper fax, the Xerox 3006 sends pages from its automatic feeder at a rate of six seconds each. It includes a host of special features, such as contrast control, confidential memory, delayed transmission, mailboxes, and multilingual capacity. As a fax receiver, printer, or copier, it provides laser quality prints, reduces and enlarges, and collates documents automatically. All this capability is contained in a unit that weighs just over eight kilograms. See Figure 2-2.

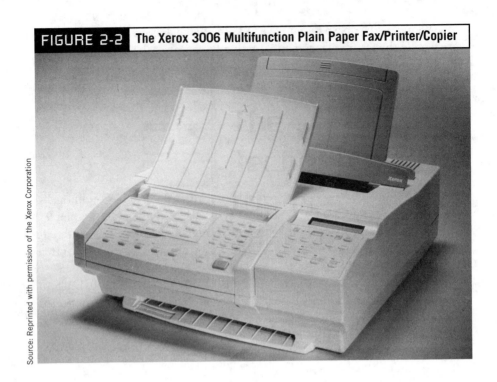

FIGURE 2-2 | The Xerox 3006 Multifunction Plain Paper Fax/Printer/Copier

Source: Reprinted with permission of the Xerox Corporation

Desktop Publishing

In the very recent past, the publishing task was long and complicated. It began with a manuscript that was sent to an editor for changes. The author then approved or disapproved the changes, revised the manuscript, and resubmitted it to the editor, who then sent it out to be typeset. Type compositors typeset the copy and passed it on to a proofreader, who checked the typeset copy to make sure it agreed with the manuscript. Next, if the offset printing process was used, the typeset copy was photographed, and offset plates were made from the negatives. If the type was "hot," (linotype or monotype), it was sent instead to a foundry where printing "mats," which would go onto a printing press, were made. Following all of these steps, the pages were printed, collated, and bound into copies of the final document.

Desktop publishing permits all of these operations to be handled, quite literally, from the desk of the author or editor. The editing job can be done right on the personal computer using word processing software. Large blocks of material can be "cut and pasted" electronically, most grammatical and spelling errors can be caught and corrected, and the manuscript can be converted to a technical document or book without its ever having been printed on paper before.

Many firms have facilities for producing camera-ready copy with laser printers or other high-quality printing devices connected directly to personal computers. Operators of this equipment are able to produce excellent page layouts, including most drawings.

Although planning and coordination are still necessary when desktop publishing equipment is used, some of the time and effort that would otherwise be needed can be eliminated. If you are fortunate enough to work for a firm that has such facilities, learn all you can about them.

Knowledge of the entire document production process is essential for planning any writing job, no matter what process is used. The following paragraphs describe the steps and equipment typically necessary for turning out a job by the desktop publishing process.

Word Processors

KINDS OF WORD PROCESSORS

Many kinds and brands of word processing hardware and software are available, and each has its advantages and disadvantages. Firms that have large mainframe computers established for purposes other than word processing and publishing sometimes acquire the software needed to include word processing as a capability. Unfortunately, these "add-on" features are not as versatile as the other products now on the market. Personal computers (PCs), which can be arranged in local area networks (LANs) with access capability to other networks and mainframes, are now common in modern offices. A variety of excellent word processing and graphics software is available for PCs, which can feed data directly into automatic typesetting systems or, using sophisticated computer printers, produce camera-ready copy for offset printing or xerography.

A WORD PROCESSOR CAN IMPROVE YOUR WRITING

A word processor will not automatically help you become a good writer. In fact, some people maintain that the word processor is a curse because it can fool its users into believing they are producing good writing when they are not. Unfortunately, there is some truth to that claim. Beautifully printed documents that contain all sorts of graphic enhancements are sometimes poorly written. Even when authors diligently use their built-in spell checkers, misspelled or poorly chosen words remain. A spell checker does not tell you when you have chosen the wrong word; it only identifies letter groups that do not spell words. Therefore, if you use "affect" when you meant "effect," for example, the spell checker will not call your attention to the error.

Grammar checkers present an even greater danger because they do not check grammar at all; they check usage taboos, most of which are better left unobserved (see Chapter 6).

Apart from all these shortcomings, however, the word processor can help you to learn to write. One of its most powerful characteristics is that it lets you see and consider your thoughts. In pre-computer times, writers had to organize their thoughts before committing them to paper, and the second draft was usually the final draft because changes beyond that point would most often require yet another entire retyping. The word processor has eliminated that problem by removing the need to commit anything to paper before the writer is completely satisfied with it. If you need three, four, or ten drafts, only your own time and effort increases. When you make an addition to a page, the word processor simply shifts all that follows forward. Of course, if you write ten drafts of a document, you will have spent a lot more time than you would in writing one or two. And that is how your word processor will help you learn to write, silently encouraging you to read and reread what you have written before you print it.

Computer Printers

Printers that can produce camera-ready copy (for the offset printing process) directly from computer disks are available from a variety of manufacturers. Many of these printers use a laser process in transferring the image from the electronic storage medium to the paper. Even color printers are now available at a reasonable cost. The Xerox XPrint 4915 Color Laser Printer, for example, can make print quality black and white copies at 12 pages a minute and extremely high-quality, full-color copies at a rate of three pages a minute. See Figure 2-3.

Reproduction

The desktop process can end with the electronic copy, which is then used to set type, or it can lead to camera-ready copy that can be reproduced in a va-

FIGURE 2-3 | Xerox XPrint 4915 Color Laser Printer

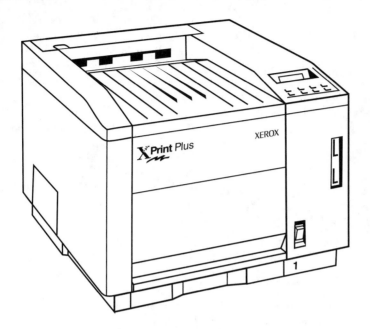

riety of ways. Photocopiers have become so sophisticated in recent years that the reproduction and finishing of copies can be done right in the office. These machines can control the entire process in a way that writers and publishers in previous decades never even dreamed possible.

The Xerox 5390, typical of this new generation of photcopiers, is capable of producing bound books with covers and even tabs or dividers. The quality is comparable to metal-master work done on offset duplicators. Text, line art, halftones, and even unscreened photos can be reproduced with high-quality results. The 5390 turns out 1- or 2-sided copies from 1- or 2-sided originals. Each chapter or section can be made to automatically start on a right-hand page. Image sizes can be varied from 64 percent to 155 percent in 1 percent increments.

Books from 15 to 125 pages thick can be permanently bound with a thermal adhesive tape binding, which is available in various colors to coordinate with cover stocks, tabs, and dividers. On-line stitching is also available, both single and double, for sets between 2 and 70 sheets. See Figure 2-4.

Source: Reprinted with permission of the Xerox Corporation

FIGURE 2-4 The Xerox 5390 Photocopier

The World Wide Web

Almost everywhere these days we hear people talking about "the Internet," "the Web," or "the World Wide Web," and with good reason: never in human history have such vast quantities of information on every possible subject been so readily available, literally at our fingertips.

A word of explanation is needed to clarify the relationships among the terms Internet, Web, and World Wide Web. Internet (actually, interconnected networks) is a generic term used to describe the worldwide system of communicating computer networks. The Web, or World Wide Web (the terms are interchangeable), is a part of the Internet, albeit an already huge and rapidly growing part, distinguished by its capability for transmitting and storing graphical information, for example, motion pictures, video, photographs, drawings, and even famous works of art.

By far the most often used feature of the Internet is e-mail. Anyone who has an Internet connection can communicate almost instantly with anyone else anywhere in the world. Most people familiar with e-mail agree that it offers a great advantage over traditional mail, which is derisively called "snail mail" by many Internet users.

The growth in the popularity of the Internet over the last few years is probably attributable mostly to the Web's graphic capabilities. Anyone with the right computer setup can make information available to everyone else on the World Wide Web. The information can be nearly anything one could imagine: product advertising, essays on a favorite topic, travel guides, opinion surveys, sales orders, or even autobiographies. The only requirement is that information destined for the Web must be cast in a special format, since computers are not yet smart enough to read or understand human languages.

That special format is called HTML, or Hypertext Markup Language. The responsibility for preparing HTML documentation often falls to technical writers. So at least a basic knowledge of HTML should be part of the repertoire of writers undertaking Web page design. (See the Appendix for an instructional exercise that contains enough information to allow you to construct a simple Web page.)

Elaborate Web page designs require intricate HTML scripts, often demanding skills beyond those possessed by most technical writers, or by most other computer users for that matter. The software industry, ever vigilant, has developed a number of HTML authoring tools that can automatically transform most word processing documents into HTML. But automation notwithstanding, knowledge of what is taking place behind the scenes in the HTML transformation process is of value to the modern technical writer. A computer-generated HTML script nearly always needs some human intervention for completion of an acceptable Web page design.

Besides learning the basics of HTML, technical writers should become conversant with Web technology and all its amazing features. However, because a detailed discussion of the World Wide Web would take at least one very large book, no detailed treatment is attempted here. Readers can consult any bookstore for a wide range of texts about the Web or, appropriately, consult the Web itself for exhaustive citations about the Web and related information.

CHAPTER 3

Team Writing

OBJECTIVES: In this chapter, you will discover that major technical documents are nearly always written by groups of people, each of whom serves one or more functions in the group. You will also learn about the specific responsibilities of team members serving different functions and about techniques for working together.

Technical Writing Groups

General Information

Major technical documents are almost always written by more than one person. The reasons are simple enough. The expertise needed to produce a complex technical document hardly ever resides with one person. And because most technical documents are produced on a very tight schedule, no one person could ever get the work done on time.

Politics, as the old saying goes, makes for strange bedfellows, and technical writing works pretty much the same way. An expert in a field is, by definition, one who has a particular interest and focus. Sometimes, people with narrow interests have a hard time seeing the importance of areas of interest other than their own. Moreover, one or more technical experts in a writing group will, more likely than not, be unable to appreciate the need for deadlines, textual balance, brevity, clarity, and so on.

It follows, then, that the word *team*, used in phrases describing technical writing groups, does not always mean that every member is inclined to teamwork. Even though technical writing groups may be made up of people preeminent in their fields, they are not teams in the same way that winners of a World Series or the Stanley Cup are. Rather, because of their *ad hoc* nature, technical writing teams can disappoint in the same way all-star teams sometimes do. That is to say, team members may perform well individually but function poorly together as a team.

Ad Hoc Groups

Some companies have major publications departments with specialists in technical writing, editing, technical illustration, publication production, printing, and graphic reproduction. These departments often operate on their own budgets and submit separate bids on the technical documentation within a contract. In many other companies, however, an *ad hoc* group is assembled to produce a specific document and is disbanded when the job is done.

In companies with a narrow scope of products or services, groups may be composed of the same people time after time, but even these groups are *ad hoc*, which means that they come together when a technical writing job needs doing and go back to their other jobs afterwards. In companies with diverse services or product lines, the *ad hoc* technical writing group typically includes one or two core members and at least one member from each of the technical groups directly involved.

For example, if a company that manufactures sensing devices, avionics equipment, electronic signal processors, commercial electronics, and computer-controlled robots decides to write a proposal for a system to map a section of the ocean floor, it will need contributions from experts in sensing devices and signal processing. When the same company wants to write a proposal to automate a manufacturing process, its proposal team will need experts in signal processing and industrial robot design.

In either of the foregoing examples, the group formed may be led by a technical writer, the technical experts acting as contributors to the final document. In small companies, which often don't have technical writers, an administrator or one of the technical experts might lead the group. In either case, someone has to set a schedule, and someone (often the same person) has to be the driving force.

A general truth about writing groups is that the likelihood of success is directly proportional to the group leader's knowledge of the writing process and that person's ability to direct the efforts of each group member while maintaining the schedule.

Group Dynamics and the Writing Process

The document schedule should take account of the steps that every technical document must go through (see Chapter 6). These steps dictate the need for a variety of skills. Before going on to the discussion of the various functions within a group, you will benefit from considering a few definitions and distinctions.

What Is Writing?

An expert in aerodynamics wrote a draft describing the movement of air over a particular airfoil. The draft was filled with grammatical errors, which compromised logic and created ambiguity and uncertainty. A technical writer who had some knowledge of the field, though not nearly as much as the expert, used this draft to produce an understandable description of the principles involved. After the piece was printed in a technical document, the aerodynamics expert showed it to everyone and claimed to have written it.

In fact, the technical expert had not written it; the technical writer had. This particular expert thought he had written the piece because he was able to understand both his own draft (*he* knew what he had intended to say) and the technical writer's version. Therefore, he incorrectly concluded that the two pieces of writing said pretty much the same thing. The truth of the matter is that they were *essentially* different from one another. The word *essentially* is used in the preceding sentence to refer to the essence (called "such-being" or "real substance" by some philosophers) of this piece of writing. The distinction may appear trivial at first glance, but its importance will become apparent when you fully understand the writing process, discussed in Chapter 6.

What Is Editing?

The jobs of the technical editor and technical writer are sometimes confused. Unlike the technical writer, the technical editor makes only non-substantial changes to a draft, being careful to retain both the author's message and style.

What Is Proofreading?

Finally, there is proofreading, the purpose of which is only to make sure that a manuscript and a typed or typeset copy (called a proof copy) match each other exactly. Technical writers must know about these distinctions and use the terms correctly in order to maintain a clear view of the writing process.

The Make-Up of a Technical Writing Group

As the foregoing discussion implies, many functions need to be filled within an *ad hoc* technical writing group. These functions include the following:

Group Leader

Scheduler

Technical Expert(s)

Editor

Production/Reproduction Coordinator

Often, several jobs are done by one person. If the group leader is a technical writer, for example, this person may schedule the project, rewrite or edit the copy, and coordinate the production of the final product. The key to a successful group is, in all cases, the group leader.

Obviously, the ideal person to lead a technical writing group is a technical writer. If a technical writer is not available, someone who at least understands what a technical document is and how it reaches its final form should lead the group. Good intentions alone will not result in a professionally produced technical document. Only an application of the knowledge of the processes of writing, producing, and reproducing technical documents, along with a good measure of administrative skill, can do that.

Successful group leaders begin by considering the desired end product and then consulting the people who will perform each step along the way. Group leaders must realize that planning, writing, and editing make up only the first half of a writing project and that production and reproduction make up the rest (see Chapter 7). Figure 3-1 shows a typical work flow diagram.

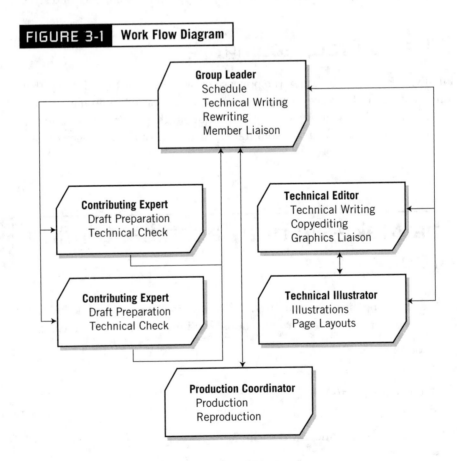

FIGURE 3-1 Work Flow Diagram

The Group Leader

SCHEDULES

The schedule can make or break any writing job, but especially one that will be done by a group. The best way to make a schedule is to start at the finish point, allotting the time needed for each step and going back through the project to the beginning. The following list shows the time line and completion dates for all steps in writing an interim report to be submitted as a line item of a contract (see Chapter 10).

May 1	Twelve copies of First Interim Report, Line Item 5, Contract 91935, due at customer's facility
May 1	Twelve copies of report received at customer's facility
April 30	Twelve copies of report shipped to customer's facility by overnight courier
April 29	Final check of printed copy before binding
April 27–28	Weekend
April 22	Camera-ready copy to print shop
April 22	Final corrections to camera-ready copy if required
April 20–21	Weekend
April 15–19	Final editing, last-minute changes, and approvals; copy laser printed to make camera-ready copy
April 13–14	Weekend
April 11–12	Edited disks checked; illustrations inserted; page layouts complete
April 8–9	Changes requested by technical contributors made and final approvals obtained of all copy and illustrations
March 22–April 8	Technical writing/rewriting and editing of technical contributions; all graphic illustration assigned to technical illustrator
March 19	Deadline for drafts from all technical contributors
March 8	Outline approval and assignments to technical contributors
March 5	Outline preparation and briefing of team members

As shown in the list, nearly two months have been allotted for writing and producing this report. Having that much time to write a report results when contract administrators have had the foresight to request that the specified reporting dates follow the quarterly intervals. Assuming that the contract runs from January 1 to December 31 of the same year, the quarterly intervals will be April 1, July 1, and October 1.

Setting the report dates back one month (to May 1, August 1, and November 1) means that the reporting team will be able to begin its work when the quarter ends, or as given in the example schedule, even a month before it ends. When a report is due the same day the quarter ends, the time line for the reporting team is much shorter, last-minute changes are common, and document quality often suffers. Also, the cut-off date for reporting progress usually has to be set back a month.

Some projects include monthly reports, which are usually prepared in a letter format. In such cases, the monthly reports form a basis for the quarterly reports and shorten the time needed for their preparation. Likewise, the quarterly reports are very useful to the team writing a final report.

TECHNICAL WRITING/REWRITING

A technical document comes to life during this step. The group leader, who either does the work or assigns it, must make sure that the rough notes and sketched-out thoughts of technical contributors are brought together into a cohesive, informative, economical, and clear expression of ideas and information that will fulfill the purpose of the communication.

Group leaders who do the technical writing/rewriting job themselves have an advantage over those to whom they may assign the step. A group leader who has questions about a contribution can approach the contributor from a position of authority, whereas the assigned writer may have to appeal to the group leader if a contributor refuses to cooperate. Such problems are discussed further in the following paragraphs.

MEMBER LIAISON

Diplomacy and Authority

In directing team members, the best group leaders combine diplomacy and authority. Diplomacy, which is neither more nor less than courteous behavior, flows naturally from a genuine respect for all members of the group. At times, group leaders find it necessary to exercise their authority, and although they should remain courteous and respectful when doing so, they may have to be firm. Group leaders must remember that their first purpose is not to make the group members feel good about themselves or to make them like the group leader or each other. "Soft-side" goals such as these cannot be achieved directly anyway. And when such goals replace real ones, the group will not produce its best work. In the final accounting, however, good work can bring self-esteem and camaraderie to a group's members.

Conflict Resolution

Group leaders often have to deal with conflicts that arise between group members. When members cannot work together, sometimes they can nevertheless contribute separately without consulting one another. However, the group leader must first thoroughly understand the problem. Conflict resolution consists in correctly identifying the nature of the conflict. After doing so, the group leader may reshuffle responsibilities, direct the parties to perform their specific jobs in a particular way, replace one or more members, or apply a combination of these remedies.

Handling member conflicts is a simple matter if the group leader understands that the issue dividing the members, not the personalities of those involved, is of paramount importance. Failing to identify the nature of a conflict can destroy a project. Consider the action taken by the group leader in the following example.

The Hospital Construction Proposal Group

Barney Kennechuk, a newly graduated, 22-year-old technical writer, has just recently joined Big Eagle General Construction, Inc. His first assignment is to lead a technical writing group charged with writing a proposal for the construction of a new hospital. He has never before worked with the people making up the group, and he soon discovers that his medical expert, Dr. John Painway, and the chief construction engineer, Ms. Louise Earl, are in conflict. Dr. Painway and Ms. Earl are supposed to work together to produce a draft for the design of the surgical suites. When the due date for contributions from all the experts is only two days away, Barney receives a memo from Ms. Earl, who writes that she is proceeding with the draft alone, since Dr. Painway has been unable to offer any workable suggestions, and that she will have it done on time. Barney notes that Ms. Earl has sent a copy of the memo to Dr. Painway.

Barney knows that Dr. Painway is a consultant who was hired specifically for this project and that his office is on the floor below his own. Barney recalls that Dr. Painway was attentive and seemed eager to get on with the job during the initial meeting to assign the various tasks. He also remembers

continued →

that Ms. Earl was businesslike and very formal. He knows also that Ms. Earl has worked for Big Eagle for 20 years, the last 12 of them as chief construction engineer. Although Barney has never visited Ms. Earl's office, he knows that it is just down the hall from his own. Barney decides to see Ms. Earl first since she is the one who indicated that a problem exists.

When Barney goes to Ms. Earl's office to discuss the problem, she is courteous, but she is also very firm. She tells him that Dr. Painway has no conception of the limitations imposed by engineering standards and that when she tried to explain to him why his ideas for the construction of the surgical suites couldn't be realized he didn't appear to understand. Seeing that the new technical writer is slightly taken aback by her direct manner, Ms. Earl softens her tone a little, saying she knows she may appear to be exaggerating, but that those are the unadorned facts. She adds that Barney should not worry and that, as stated in her memo, she will deliver the completed draft to him on the due date.

Barney Kennechuk thanks Ms. Earl, leaves, and takes the elevator to the floor below where Dr. Painway's office is located. Dr. Painway, who is confident and poised, says that he has done everything in his power to defer to Ms. Earl within her area of expertise, but that as a surgeon he has had to put his foot down about certain features of the surgical suites. He adds that he has never before encountered anyone quite so difficult to work with. As Barney rises to leave, Dr. Painway smiles and says in a tone that suggests he is sharing a secret with someone in his own age group that Ms. Earl probably has some pretty old-fashioned ideas. Barney is impressed both by Dr. Painway's manner and his calm assessment. He tells him not to worry, that he intends to take charge of the draft, and that all will be well.

Satisfied that the problem is nothing more than a clash of opposite personalities, Barney returns to his office. Two days later, Ms. Earl delivers her draft as promised. As soon as she leaves, Barney takes it down to Dr. Painway and tells him to go ahead with any changes he thinks necessary. He tells him that he will extend the deadline for two days so that his input can be included.

When Dr. Painway returns the draft not two but four days later, Barney sees that he has drastically altered it and has even redrawn the mechanical and wiring diagrams. Now four days behind schedule, Barney can't return the altered materials to Ms. Earl for her approval, and so he sends the illustrations to Bridget Nicholas for final preparation and the text to Maura La Rosa for editing and production layout, explaining to both of them that overtime work will now be necessary to get the project back on schedule.

The proposal is finished and delivered on time.

In the foregoing example, the group leader believes he has solved the problem created by the conflict. But has he? Can the surgical suites be built according to the proposal? Will the experts who examine the proposals submitted by all the firms competing for this contract be able to distinguish between good and bad engineering? The answers to the questions are no, no, and yes.

This proposal is doomed to failure because Barney Kennechuk decided on the basis of a few brief interviews that the conflict between his group members had grown out of diverse personalities. Had he tested his assumption, he would have found that Louise Earl was a nationally known engineer with more than 30 years of experience in building all kinds of structures, including no fewer than eight major hospitals across the country. He would also have discovered that Dr. Painway had moved to the city just a year earlier from a small town where he had been banned from practicing at the local hospital, that four malpractice suits were currently pending against him, that since moving he had been denied hospital privileges in every hospital in the city, and that the medical association was currently reviewing his qualifications.

The point to be noted here, however, is *not* that Barney Kennechuk should have taken the time to assemble information to test his assumption. The point is that *he should never have made the assumption in the first place*. In so doing, he ignored the possibility, which was in this case nearly certain, that one of these two group members was right and the other was wrong. He doomed the project to failure by assuming that the dispute was just a clash of diverse personalities and that it had no real substance.

Group leaders should always follow five steps in resolving conflicts.

Five Steps to Conflict Resolution

1. Always assume that disputes are about real issues.
2. Review the issues in a meeting of all parties in the dispute.
3. Discover the "object" of the dispute and cast it into a proposition that can be determined as either true or false.
4. Establish the truth or falseness of the proposition. Call in other experts as required.
5. Take appropriate action (direct the parties to work in isolation; re-assign responsibilities as needed; remove or replace members from the team as required).

In the example, the medical expert said, "Do this." When the engineer said, "It can't be done," the doctor insisted that it would be done. Who was right? Obviously, both of them weren't. The group leader's assumption that his team members had a personality conflict prevented him from seeing and solving the real problem, which he should have stated as follows:

The surgical suites can be built in accordance with Dr. Painway's specifications. Is this proposition true or false?

The proposition is false, but Barney fell into a trap that more experienced administrators avoid. The following paragraphs discuss this phenomenon in some detail.

Failing to follow Step 3 above results when the group leader adopts a very common, but deadly, mode of thought. When considering a proposition, non-objective thinkers ask the wrong question. Instead of casting the "object" of the dispute into a proposition that can be judged as true or false, they ask: *Why would people behave this way?* Looking at conflict problems from this view-point, one is naturally distracted from thinking about the character of the issue. This kind of logical error is called a Bulverism, after C. S. Lewis's fictional character, Ezekiel Bulver.

Bulverists believe, often subconsciously, that objective truth does not exist and that all disputes are about personal opinions and preferences. This way of thinking can only cause grief in the real world. Suppose that two people are working on a mission proposal to send astronauts to Mars. One of them writes in a draft that Mars has a sidereal orbital period of 686.98 days and that its min-imum distance from the earth is about 55,681,800 kilometres. The other insists that the draft needs to be changed because the Martian year is exactly the same as that of earth and that Mars is only 34,000,000 kilometres away. Through a series of clinical examinations, a competent sociologist, psychologist, or psychiatrist *might* be able to explain why those two people hold such di-vergent views. But a technical writer leading a group preparing a proposal for a mission to Mars had better leave psychology to specialists in that field and find out who is right about that planet's orbital period and minimum dis-tance from earth.

The group leader does not need to know *why* the two people hold divergent views. To get on with the job, the group leader just needs to know which one of them is right. The trap that technical writing group leaders must avoid, then, is to remember that they are not sociologists, psychologists, or psychi-atrists. When group leaders trying to resolve conflicts ask the wrong ques-tions, they court disaster.

In summary, the technical writing group leader's job is not to make peo-ple feel warm and fuzzy; it is to get a well-written technical document produced on time. Achieving that purpose does not imply tactless or abrasive behav-ior. Just the opposite is true. One need not be a genius to know that assuming psychological or social reasons for member conflicts in a technical writing group is an improper step in conflict resolution.

The Technical Editor

As mentioned earlier in this chapter, editors must respect the author's writing style or "voice" as it is sometimes called, but they also can and should suggest stylistic changes that they think will improve the copy. Copyediting cannot cure the diseases wrought by imperfect organization and grammatical structure. In simpler terms, copyediting cannot create the message.

So what can copyediting do? Grammar, logic, spelling, rhetoric, style, etc., are not equal in importance. While copyediting can "clean up" minor grammatical errors and correct misspelled words, it cannot magically convert poor grammatical structure or illogical organization into a coherent whole. That job has to have been done by a writer before any editing takes place. However, in addition to cleaning up minor grammatical or usage errors and fixing spelling mistakes, the copyeditor can (if the author agrees) change and improve the writing style, too, thereby enhancing (not creating) the message.

Suppose that you, acting as a technical editor, are asked to edit the following paragraph that is to appear in a technical report:

> The production model of the Griffin feature the UA(2)-180 wing and three engine options: a 100 hp Continental 0-200, a 105 hp Subaru conversion, and an 80 hp Suzuki Turbo conversion. At 12,000 ft, the Suzuki Turbo will pull the Griffin along at a 140 mph clip. Stall speeds for both the Continental and Subaru are 48 mph at full utility gross of 1500 pounds. With the Suzuki engine and a stall speed of 45 mph at 1200 pounds, the Griffin qualifies as an advansed ultralight airplane in Canada.

As a copyeditor, you would notice that the verb *feature* in the first line should have been *features* to agree with its singular subject *model*. In the last line, you would also correct the spelling of *advansed* to *advanced*. You would also suggest to the author that some of the advertising-style language be changed to suit the medium of a technical report. The boldface type in the following version shows the stylistic changes you might suggest.

> The production model of the Griffin features the UA(2)-180 wing and three engine options: a 100 hp Continental 0-200, a 105 hp Subaru conversion, and an 80 hp Suzuki Turbo conversion. At 12,000 ft, the Suzuki Turbo **produces a top speed of 140 mph.** Stall speeds for both the Continental and Subaru are 48 mph at **the** full utility gross **weight** of 1500 pounds. With the Suzuki engine **installed, the** stall speed **is 45 mph** at 1200 pounds of **gross weight. In this configuration,** the Griffin qualifies **in Canada** as an advanced ultralight airplane.

As you can see, these changes do not create the message, but they change the writing style of the original, which is perfectly suitable for a news release or magazine advertisement, into a style one would more likely encounter in a technical report.

The editing step should produce a writing style that is fairly consistent from one contributor to the next. But a so-called "seamless" writing style, that is, one that makes it impossible to tell who wrote what, is not a practical objective. Most of the people who read a technical document will understand that more than one person has written it and will not be at all disconcerted by writing styles that vary within the confines of technical prose.

The Contributing Technical Expert

Contributing technical experts must make their contributions on schedule and in language clear enough to convey the message. Often, clarity is not present in an expert's draft, and experts must be willing to take the time to explain concepts that they might have written in unclear language. They will be patient with writers who do not occupy their level of expertise if they realize that the expertise of many readers will not match theirs either.

Experts also need to provide references to illustrations within the text and to sketch illustrations in enough detail for a technical illustrator to make a final drawing. Although the technical writer and editor will make sure that illustrations and text match up, the expert can help by paying attention to such matters during the writing of the draft.

Finally, since technical experts are ultimately responsible for technical accuracy, they must perform a final technical check of their contributions.

The Technical Illustrator

The technical illustrator must work closely with whoever is responsible for coordinating the artwork and must maintain liaison with each person submitting sketches that will be converted to illustrations for the final document. Working with the production coordinator, the technical illustrator may also be responsible for combining the illustrations and text to make reproducible copy for the offset printing process. However, this step is now often done on a computer screen, and the illustrator's scissors and glue pot will eventually disappear.

The Production Coordinator

The production coordinator needs to make certain that all the parts of the document are in the proper sequence, that the artwork is correctly placed, and that enough copies are reproduced to satisfy the distribution list. The production coordinator and often other members of the group, including the group leader, will perform final checks to make sure that each page of the document has been properly set up for reproduction and that each copy of the reproduced document has been correctly collated and bound.

Exercises

3-1. Forming a Technical Writing Group

Note: The purpose of the exercise is only to plan for a team effort. However, at your instructor's direction, you may complete Chapters 6 and 10 so that your team can produce the report.

1. Decide what kind of a team you would assemble to write a report on the effectiveness of the student rules policy at your school. Would technical experts be able to help? Can you think of any examples that might help you to recommend changes to the rules?

2. Plan the individual assignments that you, as a group leader, might make in writing the report, assuming that contributions from technical experts could be edited and used to support any recommendations you come up with.

3-2. Resolving a Conflict

Read the conflict scenario below and answer the questions that follow.

Two members of a technical writing group that you have been appointed to lead have come into conflict and now refuse to speak to one another. Doris Mroczek is a writer, and Farley Devonshire, a technical expert. You assigned Farley the job of generating a first draft that Doris could turn into a professional-looking report. But because Farley was unable to generate a first draft, Doris attempted to salvage the job by writing a first draft and giving a printed-out copy to Farley for comment. Instead of using the copy as a basis for developing the topic further, Farley scribbles comments, some of them illegible, and sarcastic questions all over the printed-out copy. On the final page of the draft, he has printed the words "UNACCEPTABLE TRASH!" in huge block letters.

By making a few discreet inquiries, you discover that part of the problem is that Farley Devonshire, a new employee, doesn't know how to use word processing software and refuses to learn. You suspect one or both of two possibilities, each one far more serious than Farley's lack of skill with word processing: Farley is unwilling to make the effort needed to formulate his technical knowledge in writing; Farley doesn't have the required technical knowledge.

1. After reviewing the five steps to conflict resolution given in the chapter, write out an agenda for a meeting to resolve the conflict.

2 After deciding whom to invite to the meeting, write out the specific questions you would ask each person attending. Remember that your purpose is to discover the object of the dispute and to cast it into a proposition that can be tested.

3-3. *Editing Technical Prose*

Edit the following selection. Make only those changes essential to correct spelling and usage. In other words, respect the author's style, and do not rewrite the selection.

MEMORANDUM

To: All Pellston Division Managers
Date: June 22, 1998
From: A.J. Monroe, President
Subject: Change to Meetings Policy within Divisions

Meetings cost a lot of money, and before calling one, Pellston division managers need to ask themselves the following question: Is this meeting really necessary?

If ten people attend a 2-hour meeting, more then just 20 hours of employee time are invested. The time spent in preparation and travel to and from the meeting, along with the time sacrifised to lost productivity, must also be factored in. Meetings are for exchanging information that can't be transferred in some other, more efficient way. A meeting held for any other purpose simply gets in the way of productivity.

We must especially gaurd against using meetings to justify bad policy decisions, which everyone makes now and then. Airing your employees' grievances is not dealing with them, even though letting people have their say in an open meeting may create the illusion that you have done something positive. Such meetings accomplish little or no real work, and again, waste employee resources. Last year, no fewer than 17 such meetings were held in our eight company divisions and consumed nearly 5600 hours of employee time. Moreover, the loss does not take account of the profit these employees may have generated had they been free to do their jobs instead of wasting time preparing for and attending meetings. If you've made a bad policy decision, you probably know it better than anybody else. Bite the bullet, and change the policy. Don't hold a meeting about it!

For the foregoing reasons, the new Pellston policy is to scrub all regularly scheduled meetings and to call meetings only on an *ad hoc* basis. Guidlines governing the calling of *ad hoc* meetings will be issued next week.

SECTION II

The Process

CHAPTER 4

Research and Documentation

OBJECTIVES: Since the subject matter that technical writers deal with usually relates to a unique product or service, the technical writer's approach to research is somewhat different from that of the academic writer. In this chapter, you will learn where to find information; what documents to consult during the preplanning stage of a technical writing project; and which methods to use when preparing questionnaires. The chapter also makes clear that technical writing ethics demand that writers credit their sources of information, and shows you the documentation systems from which to choose when citing sources.

Research

The Contract and Other Relevant Documents

When you are faced with writing a technical document, you will want to begin at the beginning, and the beginning is the authority that says a document is needed. In many cases, that authority will be the contract. Although a contract will not tell you much more than the date the document is due and the number of copies that have to be submitted, it will often refer you to relevant specifications, which in turn refer you to still more relevant specifications. You will learn more about specifications in Chapter 7.

Other pertinent documents include those that the company has previously submitted on the same or similar projects. Unless you work for a company that has never submitted a proposal, a contractual report, an operating manual, etc. (or you have started your own new company), you will be able to find many models on which to base the technical document you are to write. The one pitfall in this kind of "research" is that you may be inclined to repeat the mistakes of those who preceded you. Therefore, you should regard these documents as secondary authorities only. When the specifications governing the document you must write confuse you or contradict one another (as sometimes happens), having the work of your predecessors at hand is a definite advantage.

Interviewing Sources

Occasionally, you may need to collect data from several, or even many, people. Since it may not be practical to visit all of them, you may wish to prepare a questionnaire. Even if you plan to interview your sources in person, you should write out the questions that need answers before going to the interview.

Preparing a questionnaire is a simple matter if you follow three simple rules:

1. Write clear questions that ask what you want to know.
2. Leave your biases out of the questions.
3. Keep the questions and the questionnaire short.

You must carefully analyze the situation before putting together a questionnaire. Sending out a sloppily prepared questionnaire is just like going unprepared to class. Precisely what do you want to find out? Do you want information, do you want opinion, or do you want confirmation that your assessment of a situation is the correct one?

If at all possible, *questions seeking information* should be written in such a way that a *yes* or *no* answer can be given. In complex situations, a question that offers multiple choice responses may be appropriate. When listing multiple choice responses, however, be sure to include the "other — please specify" category.

Questions seeking opinion are often given in the form of statements followed by the choices: *strongly agree, agree, disagree, strongly disagree.* You should note that *only* questions seeking opinions can be cast in this way. A question of fact cannot be established by opinion, no matter how many people may agree or disagree. For example, suppose that you wished to find out whether the North American Free Trade Agreement was good for the Canadian economy. The answer to that question resides in a state of facts, and even though the situation may be so complex that one may be hard pressed to find the answer, the belief that popular opinion can shed light on the matter is illogical.

Finally, short questionnaires will get a higher percentage of responses and are better for that reason.

Summary

Collecting relevant documents and consulting experts is an essential part of technical document preparation. A writer should begin by carefully reading the relevant parts of the contract, all specifications governing document preparation, and other pertinent project documents. Consulting with busy project personnel is best done only after one has very carefully thought about what questions need to be answered.

Sometimes a writer will cite the project documents just mentioned; for example, an interim report may be cited in a final report. Occasionally, one may also need to cite sources other than project documents in order to give credit to those who have performed original work elsewhere. When citing sources, one should follow the system preferred by the company or agency for whom the document is being prepared.

Documentation

References, Footnotes, and Bibliographies

When technical writers use the work of others, they are bound by the ethics of the profession to give credit to the originators of those works. Students often ask the following question: *What needs a reference and what doesn't?*

- All quoted materials, including illustrations and tables, need to be acknowledged. (Quotations running more than 50 words may also require permission from the copyright holder.)

- Paraphrased or summarized material that is attributable to a single source also needs to be acknowledged. However, paraphrased material that is considered to be generally known within a given field may be used without acknowledgment.

Once the decision has been made to acknowledge another's work, a second question arises: *How does one refer to the work?*

This second question does not have a single answer. In fact, there are nearly as many answers as there are business firms, professional associations, and government agencies. Before you are exposed to any reference style, you should know the distinctions between bibliographies, footnotes, and reference lists (or endnotes).

BIBLIOGRAPHIES

The term bibliography has three definitions:

1. The history or description of books and manuscripts, with notices of the editions, the dates of printing, etc.

2. A list of writings relating to a given subject or author

3. The systematic historical and technical study of writings, both manuscripts and books

You will notice that none of the foregoing definitions mentions anything about footnoting or reference systems. However, misapplication of the term *bibliography* over the years has created confusion, and some authorities now permit alphabetical lists of works to be used as a basis for reference lists.

FOOTNOTES

Citations appearing at the foot of the page on which their reference numbers or symbols appear are called footnotes. The footnote feature common to most modern word processing programs makes footnoting an easy task for writers.

LISTS OF REFERENCES (OR ENDNOTES)

Lists of references, or endnotes as they are sometimes called, have the advantage of a numerical arrangement that can be keyed directly and sequentially into the text of the document.

If the items in an alphabetic list are to be keyed to the text by numbers, the items must, quite obviously, be numbered. This kind of a list leads to the non-sequential appearance of the number keys in the text. A method other than number keys is thus desirable for alphabetical lists, and some method of citing page numbers has also to be devised. (One of the most common student errors is to omit page numbers from the items in a reference list, thereby rendering the entries useless.)

The Modern Language Association (MLA) solved the problem associated with unnumbered lists by directing that the page number along with a word key (usually an author's name) be included at the appropriate place in the text. However, this solution produces two side effects: word keys are more obtrusive than number keys, and the page number and the remainder of the citation appear on different pages. While neither of these effects can be considered a major impediment to smooth reading, some companies believe them to be undesirable in proposals, reports, and handbooks. For these reasons, many technical writers prefer either a footnote or a numbered reference system of citation.

The second problem requiring solution is that of the selection of an acceptable entry style. This style can assume quasi-liturgical importance for reasons that will not be discussed here. It is sufficient to say that one must discover the prescribed method for any given piece of writing and stick to it.

Citation Systems

A Simplified System

The following system of citation may be used in the absence of some other prescribed method. You will note that it emphasizes simplicity. Many variations exist within numbered reference systems. For example, some require the inclusion of the names of as many as three authors, while others permit the "et al." abbreviation for any book with more than one author. Punctuation also varies a great deal from one system to another, and variations in the requirements for abbreviations, underscoring, quotation marks, and italics are common.

The simplified method given below uses only two abbreviations, those being et al. and p. The first of these, et al., means simply, "and others." You should use it to indicate that an author or authors in addition to the one named wrote the work, bearing in mind that the purpose of the citation is merely to refer the reader to the source. The second abbreviation is p., which means "page number." The only underlining required is for names of periodicals, and quotation marks and italics are not used at all. Theses and papers are treated in exactly the same way as are articles in periodicals.

1. Von Hildebrand, Dietrich. Ethics. Franciscan Herald Press, 1953. p. 145.

2. Brown, Kathleen H. Personal Finance for Canadians, 2nd ed. Prentice-Hall Canada Inc., 1984.

3. Von Hildebrand. p. 167.

4. Manchester, William. American Caesar. Little, Brown and Company, 1978. p.533.

5. Grayson, Melvin J., et al. The Disaster Lobby. Follett Publishing Company, 1973. p. 202.

6. Lukacs, John. Budapest, in Love and War. Harper's, November 1988. p. 72.

7. Collins, Sean D. Some Considerations on Symbolic Representations and Natural Language in Relation to Physics. Doctoral Thesis, L'Universite Laval, 1986. p. 324.

The MLA System

The abbreviation MLA, contained in the heading above, stands for the Modern Language Association. Complete instructions for following the MLA's prescribed bibliographic style are given in the *MLA Handbook for Writers of Research Papers, Theses and Dissertations*. The MLA system of citation, which is extremely complex, makes the following distinctions, among others:

- articles from reference works
- articles in journals with a separate sequence of page numbers for each issue

- articles in journals with continuous page numbers throughout annual volumes
- books by anonymous authors
- books by corporate authors
- books by more than three authors
- books by two or three authors
- books in a series
- books of more than one volume by one author with a single title
- books of more than one volume by one author with the volumes separately titled
- books of more than one volume with different titles and authors for each volume
- books that have been translated
- books with one author
- edited books containing selections by more than one author
- editorials and letters to the editor in daily newspapers
- books in later editions
- lectures, addresses, and publicly delivered papers
- parts of books by a single author
- parts of collections of writings by different authors
- signed articles in daily newspapers
- signed articles in weekly or monthly magazines and newspapers
- unsigned articles in daily newspapers
- unsigned articles in weekly or monthly magazines and newspapers

The following list shows a few entries listed in MLA style. Note that no page numbers are included, since the citation within the text would give the relevant page number. For this reason, this system does not require multiple entries when the work cited is referred to more than once in the text.

Brown, Kathleen H. <u>Personal Finance for Canadians</u>. 2nd ed. Scarborough: Prentice Hall Canada Inc., 1984.

Collins, C. Edward, and Hugh D. Read. <u>Plain English</u>. Scarborough: Prentice Hall Canada Inc., 1989.

Lukacs, John. "Budapest, in Love and War." <u>Harper's</u> Nov. 1988: 72+.

Manchester, William. <u>American Caesar</u>. Boston and Toronto: Little, Brown and Company, 1978.

Stroll, Avrum, and Richard H. Popkin. <u>Introduction to Philosophy</u>. 3rd ed. New York: Holt, Reinhardt and Winston, 1979.

Von Hildebrand, Dietrich. <u>Ethics</u>. Chicago: Franciscan Herald Press, 1953.

Exercises

4-1. Preparing a Questionnaire

1. Prepare a questionnaire you would send to the administrators of your student union if you were writing a pamphlet on students' rights at your institution.

2. Prepare a questionnaire you would send to the U.S. Environmental Protection Agency for some of the information you would need to write a technical assessment of the economic effects of the most recently published automobile emission standards. Prepare a second questionnaire to be sent to automobile manufacturers.

4-2. Using Bibliographical Styles

1. Using the simplified reference style given in the chapter (or a style preferred by your instructor) list six books, six periodicals, and six journal articles that pertain to your field of study.

2. List the same six books, six periodicals, and six journal articles according to the MLA reference style.

3. Obtain a copy of the *MLA Handbook for Writers of Research Papers, Theses and Dissertations* from your school library and compile a list of references containing at least one each of the subdivisions within the MLA style.

CHAPTER 5

The Technical Writer's Tool Kit

OBJECTIVES: In this chapter, you will learn how to write for readers by employing methods and modes of exposition that appeal to their senses and experience. You will also learn about the uses and misuses of readability formulas and about the importance of logic in technical documents.

Writing for a Reader

Imagine yourself about to move to a new house. You have rented a truck big enough to carry all of your furniture and other belongings, and you have enlisted some friends to help. You have the job of loading the truck, and your friends have the job of driving the truck to your new house and unloading and unpacking everything without your assistance. Now, keeping in mind your friends and the job they will face at the other end, you can put everything into boxes and carefully label the boxes as to their placement in certain rooms. Or you can put things into the boxes in some other order, such as square things with square things and round things with round things.

If you follow the latter procedure, you will end up with dishes and Frisbees in one box, books and pillows in another box, and so on. The net result of your decision might be that you will lose something in the move, including, most probably, your friends.

Writing is that way too. If you don't consider the readers' needs, you may find that they will decide their task is too great. They won't bother to "unpack" the message from the pages you write. And if one of those readers is your boss, you might have to forget about that promotion you were hoping for.

Nevertheless, bad writing is very common. It is probably most common in documents unregulated by consumer demand. The editors of popular periodicals know that obscure prose doesn't sell. If one had to puzzle over the meanings of *Reader's Digest* articles, there would not be a *Reader's Digest*. Likewise, the editors of *Flying, The Atlantic,* and *McCall's Quilting* cannot afford to publish writers who don't care about their readers. On the other hand, bad writing can pop up like weeds in the rich soil of an unregulated bureaucracy.

One may find bad writing in textbooks, chosen not by the students who have to pay for them but by teachers who need texts that deal with certain ranges of topics; in contracts laden with legalese; and in the proceedings of too many professional conferences, whose delegates desperately publish (lest they should perish) though they may have nothing of any importance to say.

Sadly, bad writing also turns up in some technical documents. Recent developments in technology, however, have made bad technical writing less prevalent than it has been in the past. Consumers demand good manuals describing ways to use their personal computers or other sophisticated gear. And of course, a healthy international competition has increased the number of well-informed managers, who have always known of the direct correlation between a company's communication practices and its competitive standing.

Some people take the position that "professional" writing is different — to the extent that the reader is secondary. This position is clearly a wrong-headed one. In fact, we can't go any further without first clearing up some of the false notions of formal and informal, and of professional and popular writing.

Formal, Informal, Professional, and Popular Writing

The following quotation, taken from a behavioral science journal, is a typical example of the kind of writing found in many professional journals.

> The hypothesis postulates that in a setting in which the costs of volunteering probably outweigh its egoistic benefits, the longevity of participation as a service volunteer will be inversely related to the extent to which the service volunteer's entry was motivated by the need for or expectation of egoistic benefits.

The first step in analyzing this quotation is to figure out what its author is trying to say. After that, perhaps we can try to understand why the author chose those particular words and that particular sentence structure. The following may be a fairly accurate translation:

> We hypothesize that people who volunteer their time in return for a feeling of pride or accomplishment withdraw their services when such feelings do not adequately compensate them for their work.

The original version contains 50 words; the revision contains 31 words. Even more dramatic is the contrast in number of syllables. The first version has 96, and the second has only 51 — a 47 percent saving. When we give our readers less work to do, they (like our friends helping us move) will stay with us for a longer time. Of course, we have to take care to give them all the information they need. Saving words at the expense of meaning is worse than using too many words. Readers would then not be required to process more words than they should, but they would be left to wonder what information had been omitted.

But what's wrong with an author's using complex sentence structure and long words? After all, this is a formal document, and it does relate the details of an important experiment, doesn't it? Shouldn't it *sound* important? Let's look at our example in more detail.

The first three words are, "The hypothesis postulates." Does a hypothesis postulate? "Postulate" means the same as "hypothesize." The author didn't want to say, "The hypothesis hypothesizes," but failed to see that "postulates" doesn't do anything to alleviate the effects of the redundant pair. So the author made the reader process seven syllables in those two words to receive a single concept. The next four words, "… that in a setting in which …" mean, in this case, "that if."

The author's reasons for using these and other expressions, such as "egoistic benefits" and "longevity of participation," might well have been to imitate other researchers in behavioral science on whose work the experiments were based. Also, the author may have believed (and been justified in believing) that this journal expected this sort of writing style.

Our purpose here is not to criticize this particular author, but to point out that this article could have been written so that anyone could understand it. A simple, straightforward writing style would have permitted the author's colleagues, along with non-professionals, to readily understand the article. Would such a version be "popular" writing? Is writing popular just because it isn't in code? Is this kind of writing undesirable? The obvious answer to each of these questions is *no*.

What is lost in the second version of our example? Professionalism? Formality? The dignity of behavioral science? Most people would agree that nothing is lost. In fact, the excerpt's complex form says more about the silliness (which is definitely undignified) of some "professional" writing practices than it does about the complexity of behavioral science.

To sum up this discussion, let us say that formality is not the same thing as pomposity, and a professional approach to writing does not imply unnecessary complexity. In fact, a professional approach demands just the opposite. The technical terms and the concepts for which they stand are complex enough without an artificial complexity superimposed by a pompous style.

Readability Formulas

Many modern word processing applications contain a feature called the Fog Index, which was invented by Robert Gunning. The Fog Index uses sentence length and word complexity as criteria for judging the readability of a piece of writing. It can be very useful if it is properly understood and used as Gunning intended. Robert Gunning built his clear-writing seminars on the theme that clear writing was a product of clear thinking, and he expressed his own message with the same clarity and force he taught others to value: *If you want to be understood, write to express, not to impress.*

Gunning based the Fog Index on research that proved that even the most complex topics can be written about in straightforward language and in a simple style. And contrary to what a non-writer might expect, the length of the piece is most often reduced, not increased, by such a practice.

But readability formulas must be used properly, and their limitations must be clearly understood. The Gunning formula is composed of just two elements: percentage of polysyllabic words and sentence length.

The measurement is simply the sum of the average sentence length and the percentage of words of three syllables or more, multiplied by 0.4 to give a figure that coincides roughly with the number of years of formal education a comprehending reader will need to have completed. Some three-syllable words, such as two-syllable verbs that become three syllables when suffixes are added, are excluded from the count.

Even when accurately calculated, the Fog Index does not establish that writing that tests low is easy to read. As Robert Gunning himself often said, "The Fog Index is a handy means for judging readability. But it is not a formula for how to write ... To write clearly, you must apply the principles of clear statement."[1]

Determining Who Will Read What You Write

If you are writing a report to your supervisor, that person is the primary reader. But your supervisor is probably not the only reader. A report is filed and kept as a permanent record of the project it relates to. If this project has some significance for another project, members of that project group may read your report. If your report has significance for future financial planning, someone in the controller's office might read it. If your report says something that might be important to a future product, someone in marketing might read it. Perhaps its contents would be interesting to someone in the advertising group, who might want to use the information in a brochure.

Some of these uses might be a little far-fetched for some reports, but the point is that the broader the appeal, the better. There are, however, limits. You can't be expected to write in a style that omits technical terms, for example. By and large, you are writing for your peers, and a good question to ask yourself is, "Will this report be significantly longer if I broaden its appeal." If the answer is "yes," don't do it.

A contractual report, that is, one that is to be submitted to a customer as a line item in a contract, should appeal to as wide an audience as possible also. A customer's organization, whether it be a government agency or a private firm, will have the same array of personnel as your own company. The same requirement holds true for proposals.

---

[1] Gunning, Robert. *Reference Manual on Clear Technical Writing.* Industrial Education Institute, Boston, p. 3.

Instructions should be written so that the person who has to carry them out can understand them without any extra effort. A heavy and ponderous writing style in a manual is more than a mere annoyance; it can result in an ambiguous reading that could contribute to a disaster. Instructional writing is a special case that is dealt with in Chapter 12.

Audience Analysis and the Methods of Exposition

Technical writing is expository by nature. Although technical writers sometimes write descriptive prose, they do not (or should not) attempt to describe an object or a process without using all the available expository techniques. Technical writing modeled on old-fashioned descriptive writing practices often ends up being obscure and object-centered.

The reason for the obscurity is that the non-technical writer's methods are inappropriate in technical writing. Simile, metaphor, personification, and all the rest of the non-technical writers' tool kit enable them to engage their readers. In contrast, technical writers are denied the chance to invent fresh metaphors and, instead, often resort to vogue words, such as *viable* and *key*, in a vain attempt to breathe a little life into technical documents.

However, the technical writer has a method that most non-technical writers are denied: graphic illustration. In addition to graphic illustration, technical writers have six other expository methods at their disposal.

1. Definition   Object-centered
2. Physical Detail   Object-centered
3. Cause and Effect   Object-centered
4. Example   Reader-centered
5. Comparison and Contrast   Reader-centered
6. Analogy   Reader-centered
7. Graphic Illustration   Reader-centered

Note that the first three methods are object-centered, and the final four are reader-centered. Knowing about this difference will help you to learn to write well. Technicians, technologists, and engineers must communicate information about their work, including theories of operation; descriptions of physical equipment; arguments (sometimes not explicit) for the beginning, continuation, abandonment, or expansion of a project; instructions for operating or maintaining equipment; detailed plans for a proposed program of investigation; or a detailed account of a project already completed.

Even in purely impartial presentations, the material presented must always be clear enough so that a manager (whose technical competence may be in accounting, marketing, or some other non-engineering field) can make the right decisions. Basing technical documents solely on the object or process to be described, therefore, can easily leave most readers in the dark.

Object-Centered Methods of Exposition

DEFINITION

Most people think that definition is a pretty valuable device. "Define your terms" is a cry we often hear in the heat of an argument. While definition is a valuable way of explaining words used in a given context, it has little value as a method for describing objects and processes.

A classical definition has two parts: classification and differentiation. When we define, we classify the object defined and then attempt to exclude everything else in the class into which we have placed it. The classification we choose is, therefore, very important. If we wish to define the word *harp* and choose to classify it as *an object,* we must eliminate everything else in the universe in the differentiation stage of the definition. If instead we choose *musical instrument* as our classification, we will have to eliminate only the other musical instruments — not chickens, automobiles, books, and the Milky Way.

PHYSICAL DETAIL BEYOND DEFINITION

Definitions of physical objects will relate some physical details of the object, but usually not all of them. Therefore, this method can be used to add to the body of information if that is desirable. Physical details are, needless to say, completely object-related. Although I may choose to omit certain details, I cannot make up the characteristics of an object or a process in order to engage my audience.

CAUSE AND EFFECT

Where did the object come from? What is it made of? How did it get here? What is its significance? What can we do with it? Why do we need it? All such questions would be answered by this method. The cause-and-effect method of exposition, while it might be useful either within or beyond a definition, is nonetheless (like definition) object-centered. Just as there is a perfect definition determined not by a reader's needs but by the object defined, so cause and effect is determined by the object or process under discussion.

What would be the outcome of using only object-centered methods? The following example is a student paper with the name of the mechanism omitted. See if you can follow its logic.

A _____ is a simple mechanical device used for _____ing. The components of the _____ are the base, latch, spring, vise, and holder. The base is a flat, rectangular piece of wood approximately 10 cm by 5 cm. The purpose of the base is to serve as an attachment for the other components. The latch is a thin, rounded piece of metal about half the size of the base with a slight hook on the end. It is located at the top of the base where the end is hooked on a metal hoop that is firmly secured to the base. The latch is attached to this hoop in a chain-link fashion so that it is movable. The spring is a coiled piece of metal, running across the center of the _____ width-wise. One end of the coil attaches, pointing upwards, directly to the base. The other end rests, pointing downwards, firmly on top of the right-hand side of the vise. This arrangement gives the spring its potential energy. The vise is made up of a thin, rounded strip of metal bent into four bars. One bar runs along the bottom end, two bars run along the sides up to the center, and the other runs width-wise at the center of the base. The two bars on the side of the base are at a slant going downwards to the bottom bar. The bottom bar touches the base and is what actually effects the _____ing motion. The bar at the center runs through and is attached to the coiled spring. The holder is a small, rectangular piece of metal that lies on the lower, middle half of the base between all the bars of the vise. At the end near the center of the base is a piece of metal that is bent upwards and then slightly across at the top. This shape is to allow for attachment to the hook. At the other end, the metal is curled around with a hole at the top of it. The holder is held in place by a metal hoop running across the top of it with each side attached to the base.

The foregoing example was written as an assignment in response to a long-standing prescription for writing descriptions of processes and mechanisms; that is, the student is asked to follow a prescribed procedure in learning how to write technical material. Did you discover that the object described is a mousetrap?

First, let's look at the benefits of such an exercise. Students are forced to use only words to describe objects that they have never before given much thought to, and the experience is likely to make them appreciate the difficulty of casting physical objects in verbal form. This task will help to put students in a language mode, and they may find the exercise challenging.

However, on behalf of those students who will wonder why they have been asked to do such an exercise, let's admit that the exquisitely phrased almost 400-word description of a mousetrap is, in fact, not easy to understand. Its structure is reminiscent of leases and legal contracts, which, as we all know, often have to be sorted out by judges and juries. In other words, it will not be what is needed by someone who wants to understand how a device should be assembled or by someone who must decide whether a project should receive further funding.

So, how can we write in a way that satisfies the needs of the technician who must assemble a piece of equipment or the needs of the busy supervisor who must make business decisions based on what we write? If we shouldn't describe a lunar roving vehicle in a 400,000-word description (don't forget that it will be 1000 times more complex than the mousetrap), how should we describe it? You will discover the answers in the following discussions of reader-centered methods and modes of exposition.

Reader-Centered Methods of Exposition

EXAMPLE

An example is like a fresh breeze that suddenly blows through a stuffy room. It is a break from the mental gymnastics that the object-centered methods demand. An example permits us to "see" what our mind has been laboriously trying to construct from the words on the page. Examples are used chiefly to help a reader understand principles, definitions, or theories. They are related to "effect" in the cause-and-effect method, but they differ in that the writer may usually choose from a variety of them.

The following quotation is from an article on the importance of controlling heat in space vehicles:

> The extreme heat-flux changes that spacecraft encounter constitute a threat to payload instruments and living occupants. For example, a satellite passing into the earth's shadow receives fluxes that are ~10 times less than in direct sunlight ... .[2]

This example gives information concerning the changes in heat encountered by a spacecraft. At the same time, it gives a visual image in that the reader can mentally picture a spacecraft passing into the earth's shadow.

In a later paragraph on the same page, the authors once again give an example that provides valuable information. However, in this example, the effect

---

[2] Thostesen, T. O. and Wolf, F. N. "Thermal Control in Space Vehicles," *Nucleonics*, April 1961, p. 94.

is not visual; instead, a mention of relative amounts of solar radiation provides a mathematical comparison.

> As more powerful booster rockets are built, spacecraft will travel farther into space and encounter thermal-flux variations for longer periods of time. For example, near Mars at perihelion a 1-ft diameter vehicle receives 182 Btu/hr solar radiation; near Venus at perihelion the same craft receives 675 Btu/hr.

COMPARISON AND CONTRAST

Comparisons enable the reader to "see" the object in a familiar frame of reference. In describing a lunar roving vehicle, for example, a writer might make reference to a more familiar vehicle such as a jeep, a tractor, or a tank. (See also Modes of Exposition, discussed later in this chapter.)

Comparison must be made between like objects. However, the value of a comparison is not proportional to the number of points of similarity. If points of contrast exist, they too can be mentioned to advantage. In fact, the *raison d'être* of some comparisons is to show the differences, and thus the superiority, of one item over the other. The following comparison paragraph is augmented by an example in a second paragraph.

> Not only is the Fanwell 4-cylinder engine 80 pounds lighter than the 6-cylinder Guzzler, its output is 180 horsepower, compared to the 165-horsepower output of the Guzzler. The displacement is 360 cubic inches for the Fanwell, which burns 9 gallons per hour at 75 percent power, while the 401-cubic-inch Guzzler requires 11 gallons per hour at the same power setting.

> When both engines were mounted in identically equipped Silver Streak airplanes, the one with the Fanwell achieved a cruise speed of 138 knots, 9 knots faster than its competitor, over a measured course.

ANALOGY

Analogies compare objects or processes to other objects or processes that really bear little or no resemblance to them. As with the other reader-centered methods, an analogy's purpose is to bring the reader into contact with the subject of discussion.

A well-known analogy compares the nuclear fission process to a room full of mousetraps, each set to go off, and a layer of Ping-Pong balls covering the mousetraps (the combination representing fissionable uranium). When a Ping-Pong ball (a neutron) is thrown into the room, it sets off a few mousetraps, which bounce their Ping-Pong balls off the ceiling and walls into other mousetraps; this action sets off more mousetraps, freeing other Ping-Pong balls and creating a chain reaction.

GRAPHIC ILLUSTRATION

In non-technical writing, illustration as a method of exposition refers to a long example, graphic illustration being unavailable in most types of non-technical writing. (Chapter 8 is devoted entirely to graphic illustration.) But technical writing is in a separate category, and graphic illustration is one of the technical writer's most valuable tools, as we will see in the following discussion.

The Modes of Exposition

In using the methods of exposition, the professional technical writer, always mindful of the readers' needs, also combines the three modes of exposition: *linear*, *parallel*, and *spatial*.

The Linear Mode

Prose is linear by its very nature. That is to say, although sentences are composed of subjects, verbs, modifiers, objects, and complements, the words that constitute prose are arranged to follow one another in a line. Likewise, sentences arranged in tandem make up paragraphs, and paragraphs following one another make up technical documents. This overall arrangement of parts is what all readers expect, and the careful technical writer will never disappoint them. However, you will find that a parallel or spatial mode of exposition serves the reader's purpose much better *within* certain parts of a document.

The Parallel Mode

Comparison and contrast, one of the reader-centered methods of exposition, can often be presented in the parallel mode. Tables and certain kinds of illustrations offer the technical writer a way of placing important information side by side, thus dramatically reducing the amount of work that readers must expend in getting the significant information out of the document.

For example, a tabular presentation of the performance characteristics of two or more aircraft would permit readers to see, at a glance, a comparison of top speeds, cruise speeds, stall speeds, fuel burn, wing loading, range, empty weight, gross weight, etc.

The Spatial Mode

Do you remember the mousetrap example? One or two properly conceived and well-placed illustrations would have removed all the mystery from that tedious attempt at exposition. At the same time, graphic illustration would have reduced the number of words to about one-quarter of those in the original version. Although the *overall* scheme of every properly executed technical document is linear, technical writers must remember that objects occupying space *cannot* easily be cast into prose, which as mentioned earlier is linear by its very nature.

Considering Your Audience

Technical Terms: Yes or No?

As already stated, good technical writers make their documents appeal to as wide an audience as possible. However, one must observe certain limits, mainly having to do with technical terminology. As a writer, you cannot avoid, and should not try to avoid, the technical terms necessary for cogent discussion. You must assume that technical document users know the terms essential for discourse in a given field.

Students in a technical writing course can best practice this balance by keeping in mind that they are writing for their peers. The question to ask is, "Will Sally and Varinder, who sit across from me in my technical writing class, fully understand what I have written in this assignment?" You will soon begin to develop a sense of which technical terms are appropriate, which need some explanation, and which are inappropriate.

Applying the Modes and Methods of Exposition to Help Readers

Asking the following questions will help you determine who your readers are and what they need to know or do. You should ask these questions at every stage of document preparation and especially during the rewriting stage, which you will learn about in Chapter 6.

- Will your readers be expected to carry out a procedure? That is, have you written an instruction? What kind of instruction is it? Is it one that will require someone to enforce a company policy? If so, you may want to stay within the linear mode and give many examples of occasions that would require application of the policy.

- Will your readers be required to conduct a test of some kind? That being the case, you will no doubt want to use the spatial mode to augment parts of the procedure with illustrations that, for example, show the connections the reader will need to make between the items of equipment in the test setup.

- Will a specific member of your audience have to make a decision based on the data you present in a report? If so, considering that particular reader's needs may lead you to rely on the parallel mode and present the data in tabular form.

Logic and Exposition

Readers expect logic in the prose they read. Most of them are not consciously aware of this expectation. Logic is a natural part of our daily lives because we are rational beings. We want to know why phenomena occur and why people act or fail to act in certain situations. The structure of a well-written technical document acknowledges this reader expectation, and all good technical writers know how to apply inductive and deductive reasoning to the structure of the documents they write.

Inductive Reasoning

Inductive reasoning is the process by which we derive general propositions from specific bits of observed data. For example, if in looking around your

Although *ad misericordiam* tactics appeal to the better part of human nature, their use is dishonest — even in a worthy cause. This appeal is also very often effective in promoting unworthy causes because many people within the general public don't take the time to rigorously examine the positions of pressure groups and to formulate their own opinions on important issues.

FALSE ANALOGY

An analogy is a comparison of two things that aren't really comparable. It is quite a useful device for describing processes and theories. However, while analogy is useful as a descriptive tool, it is not a proper logical tool. Suppose that someone declares that dictatorship is better than representative government and offers the following analogy as proof.

> A ship can have only one captain, a kitchen only one chef, and a football team only one quarterback.

Thoughtful people will note that sovereign nations are much more complex and essentially different from ships, kitchens, and football teams.

False analogy appeals mostly to people who are, for all practical purposes, unaware of logic.

CONTRADICTORY PREMISES

Reasoning fails whenever it denies objective reality. This fallacy is called *contradictory premises*. A classic case of contradictory premises is to be found among those who believe that only change is real. The pre-Socratic philosopher Heraclitus was known for his observation that one could not step into the same river twice. One of his students pointed out that one could not step into the same river once, since even a single event necessarily consumed time. Whether they knew it or not, these two early thinkers had defined the philosophical limits of empirical science.

Later thinkers, such as the idealist George Berkeley, looked for a way out of this dilemma without abandoning empirical science as a philosophical foundation. Berkeley "solved" the problem by declaring that matter existed only in the mind. The Phenomenological Realists of the twentieth century really did solve the problem by distinguishing between the blunt science of empiricism and the necessary, irreducible phenomena of philosophy.

Still, some people like to say that change is the only reality and that absolutes do not exist. We might ask them *what* is changing and *who* is talking? These victims of contradictory premises believe *absolutely* that there are no absolutes.

Likewise, if we have an irresistible force, we cannot, at the same time, have an immovable object. There are no square circles or round rectangles. This principle of contradiction, although it may seem obvious to us in the examples given, is often violated.

The contradictory premises fallacy has great appeal for the anti-intellectual and for the confused and prideful reformer.

HASTY GENERALIZATION

Reasoning inductively, we must formulate our general conclusions carefully and with proper regard for the limitations of the inductive process. If we should note that in a classroom of 20 students, two students have red hair, we would not have assembled enough data to conclude that approximately 10 percent of all students have red hair. If we discover that several politicians in our government have been found guilty of crimes, we would not have enough data to conclude that all, most, or even a significant number of politicians are criminals.

Hasty generalization appeals to people who are envious, suspicious, or just mentally lazy.

FALSE DILEMMA

Occasionally, you will hear someone state a proposition in "either/or" terms: "Either we stop this tyrant here and now, or future generations will suffer the consequences." Such a formulation may be true or false, but we cannot know which it is unless we ask some questions. What are the dangers? What are the options? What is meant by the words "stop" and "consequences"? What action will we take? How will it be carried out? What will be the long-term effects of our action? Are there ethical and moral considerations?

False dilemma appeals to those who are intellectually lazy or to those who are already committed to a certain position.

STRAW MAN

This fallacy falsely restates the position it opposes (builds a straw man) and then destroys the opposing position by "knocking over the straw man." For example, let us say that Political Party A advocates reforms in the way committees and the main political body interact. The reforms will outlaw party discipline that makes it impossible for elected representatives to vote the way their constituents would have them vote.

Political Party B, left without an argument against such reforms, restates Party A's position, saying: "They want to change the methods that the founders of this great nation instituted as a check on the attempt by power-hungry robber barons to rule the country." At this point, the straw man is complete, and Party B knocks him over with the statement: "The citizens of this country won't stand for such an affront to our great heritage, and neither will we."

Party A's defense lies not in going against the "founders" in favor of the "robber barons" but in calling attention to the fact that its position has nothing whatever to do with either.

The straw man fallacy appeals to ignorance and bigotry.

Summary

Many, perhaps most, technical documents have more than one reader and more than one use. For example, one might write a memorandum report for a project engineer. But that report will become part of the permanent record of the project. In analyzing the audience for a particular technical document, the writer must constantly keep in mind the multiple uses of technical documents. Writing for a reader also implies the full use of all the methods of exposition, particularly the reader-centered methods: example, comparison and contrast, analogy, and graphic illustration. Although technical documents are arranged in linear fashion, the other two modes of exposition, namely, parallel and spatial, are also important devices that good technical writers use to help readers understand the document. Finally, technical documents must be constructed with due attention to logic.

Exercises

5-1. Writing a Paragraph Using Examples

Using examples, write a short paper on what constitutes good government.

5-2. Writing a Paragraph Using Comparison and Contrast

Compare and contrast your favorite sport with another sport.

5-3. Explaining a Theoretical Concept Using Analogy

Using analogy, explain a theoretical concept, such as the flow of electric current, the movement of weather systems, etc.

5-4. Writing a Description Using an Illustration

Describe your house or your room, providing an illustration (or just a rough sketch), in order to reduce the number of words you would have to use otherwise.

5-5. Writing a Short Paper Using Illustrations

1. Using illustrations, write a short paper comparing the performance of your car (or a friend's car) with another make and model.

2. Rewrite the mousetrap description given as an example in this chapter (p. 63); include illustrations. Compare the number of words in the original with the number required to write this description with illustrations.

5-6. Analyzing a Short Paper

Analyze an editorial or a letter to the editor in your local newspaper. Write out the syllogisms and identify the assumptions of the writer. Identify the logical fallacies.

The Steps in the Process

O B J E C T I V E S : In this chapter, you will learn how to organize and carry out the publications process, beginning with the planning of the job and ending with the final version of the document. You will learn about the value of a routine in preparing a document for any of the three major categories of technical writing. You will also discover that following this routine can help you develop your creative instincts and bring out your inherent writing ability.

Planning the Job

Technical writing falls into three general categories: reports, proposals, and instructions. Within each of these categories, however, there are subcategories and sub-subcategories. For instance, reports include contractual reports, internal reports, data sheets, capabilities statements, and other customer information literature. The subcategory of contractual reports contains interim and final reports, and that of internal reports contains memorandum reports and a host of form reports.

Fortunately, students learning about technical writing do not have to be concerned with all of these sub-subcategories. They do need to know, however, what general area each writing task falls into, and they need to know what questions to ask. The following three questions must be answered before any work can begin:

- What form will the document take?
- How many copies of the document will be required?
- When must the document be completed?

Determining the Category

The answers to all three of the foregoing questions follow from a determination of the category. For example, let's say that one is assigned the task of writing a report to satisfy the terms of a contract. One knows at once that this is a contractual report and, by referring to the specific item in the contract, can answer the first question, since the contract will specify the form of the report. The contract will also reveal how many copies must be sent to the customer. By adding the number of copies needed to satisfy the rest of the distribution list, one can answer the second question.

The contract will also specify when this report must be in the hands of the customer. The writer is then in a position to begin planning the report process. Knowing the lead time needed for the printing process and knowing the form the report is to take, the writer can establish a deadline for the writing, rewriting, and final editing. One who attempts to write a report without going through these preliminary steps may discover that insufficient lead time is available for the production and printing, or that the choice of format or illustration method does not comply with the governing specifications or style guide. (Graphic considerations are discussed in Chapter 7.)

Outlining

The most important thing to remember about outlining is that it is a process of division. Divide the topic into as many subtopics as you can think of. Some of these subtopics will be major and will be roughly parallel in importance. These major subtopics will become the primary headings of the document. Then, subdivide each one of the major subtopics. These will become the secondary headings. At this stage, place the subtopics that you didn't use as major subtopics from the first division under major subtopics as appropriate. These will become tertiary headings. A repetition of these steps for all levels of subtopic will result in a detailed outline.

The second most important thing to remember about the outline is that it is a utility item. If during the course of writing the document, you find new information or a better way of organizing the information already available, change the outline; never become its slave.

Third, use the same system of alphanumeric designation for the outline that you will use in the final document. Some alphanumeric system should be used for the outline even if one is not used in the document. (Some document styles use a system of weighted headings instead of an alphanumeric scheme. For example, primary headings may be in boldface 18-point capitals, secondary headings in 14-point boldface capitals, tertiary headings in 12-point italics, and so on.)

Finally, the outline is not necessarily the first step in the writing process, but it is an essential step. At some point, in other words, the writer must see the whole framework of the document. For some writers, this point may be reached when the document is finished! Nevertheless, the overall structure is extremely important, since it is this structure that will enable the reader to see the relationship between details.

Most people need some sort of an outline before the actual writing can begin. Such an outline may be quite simple, as in the following example:

I. INTRODUCTION

    A. Purpose of Report

    B. Scope of Report

    C. Background Information

II. THEORY OF OPERATION

    A. Electronic Circuits

    B. Hydraulic System

    C. Controls

III. PRACTICAL APPLICATIONS

    A. Environmental Control Systems

    B. Assembly Lines

    C. Efficiency Evaluation Measurement

IV. CONCLUSIONS AND RECOMMENDATIONS

A framework such as that shown in the foregoing sample outline is sufficient to provide impetus to a first draft. After the first draft has been completed, the outline can be modified as needed to provide a framework for a second draft.

This sample outline does not include front matter items, such as the abstract, the table of contents, the list of illustrations, and so on. These items, which are explained fully in Chapter 10, give the reader an overall picture of a document's structure. In order to understand how front matter performs this function, we will consider organizational possibilities in the paragraphs that follow.

Two Methods of Organization

Writers can organize material to make their messages immediately apparent, or they can try to generate interest by holding the reader in suspense. Most technical writing should give important information first. A moment of reflection will tell us that people who read technical information do not want to be held in suspense. As stated earlier, the purpose of technical writing is to inform, not to entertain.

This principle should govern the *overall* structure of most technical documents, especially reports and proposals. Most technical documents will be of interest to people with varying levels of technical expertise, and each of these people can have a different reason for being interested. The one thing they will all have in common, however, is that they all read for information.

The busy executive reads a report (or parts thereof) to help in the decision-making process that is part of an executive's job. Perhaps the marketing manager reads parts of it to get some insight into the company's prospects for movement into a new area of business. Engineers read it to understand how this development fits into something they are doing. If the report is organized in a *"top-down" development* style, each of these people will save time and effort.

"Top-down" development means that the important and significant information comes first. The front matter helps to provide for this overall structure by including an abstract on one of the very first pages, where the executive can get at it without even consulting the table of contents. (Abstracts, executive summaries, summaries, and summary digests are discussed in Chapter 10.)

The abstract may be all that this executive needs to read, since it will give the major conclusions and recommendations. After reading the abstract, marketing or contracts people may wish to continue reading, going on to the introduction and perhaps selected technical sections. Some engineers or technologists may read the entire report from beginning to end, but even they will not want to wait to get to the end of the report to discover the conclusions and their significance.

Of course, within certain sections of a document, the writer may find it necessary to develop conclusions from data in a step-by-step analysis. Or a chronological development may be needed for an adequate explanation of a certain outcome. These methods are perfectly acceptable, provided that they do not define the overall structure and do not carry on for an unreasonable length of time.

Writing the First Draft

Facing a Blank Computer Screen

The first draft should not be the hardest part of the writing task. Some people make it hard, however, by trying too much to make the writing perfect. There is an old saying among authors: *There is no such thing as good writing — only good rewriting.*

Those who try to write perfect prose in a first draft use their time inefficiently and most often achieve poor results besides. All such attempts fly in the face of the reality of the writing process, which is founded on the following fundamental truth: Words on paper form a material substance that can be considered, changed, and shaped. Ideas that remain inside your head don't have this same malleability, for the range of forms they can take is still nearly infinite.

Once you admit to yourself that your first efforts are going to fall short of perfection no matter what you do, you will have taken a big step toward making the writing process work for you. Get something on your computer screen (or on paper) as soon as possible. Don't let that blank computer screen intimidate you.

A good outline makes the first draft much easier to write than it might otherwise be. But no two people are alike in their approach to writing. Some writers prefer to think the whole project through, producing an outline in the process. Others would much rather just let the thoughts tumble out in stream-of-consciousness fashion. The beauty of word processing programs is that they take the place of the scissors-and-glue-pot method of writing.

Planning and Coordinating Illustrations and Tables

The time to plan illustrations and tables is during the writing of the first draft. Such planning will help to eliminate unnecessary writing. You will find that long and tedious explanations of processes or mechanisms will become concise and informative when illustrations are intelligently conceived and created. Likewise, lengthy comparisons are shortened by presentation in tabular form. Use the Modes of Exposition to your advantage (see Chapter 5, p. 66).

Since more than one person will write most long technical documents, the project coordinator must devise methods of unifying the document. A preliminary numbering scheme is extremely helpful to keep track of first-draft illustrations and their references, particularly if the authors are working concurrently and independently. The following system works very well:

Prel. No.	Final No.	Title	Page
HT-1		Amplifier Schematic Diagram	
HT-2		Receiver Schematic Diagram	
HT-3		Installation Methods	

The initials of the author, Harriet Thomas, are followed by a number, which is the sequence in which Harriet Thomas's illustrations will appear. However, since her contribution to the document follows Bill Stanley's and precedes Laura Lubinsky's sections, they will not be Figures 1, 2, and 3.

After the production copy is complete, each of the authors can submit preliminary figure numbers to the person coordinating the document, and that person can assign final figure numbers and page numbers based on the actual sequence in which the figures will appear. The combination of these documents will look like the following:

Prel. No.	Final No.	Title	Page
BS-1	1	System Block Diagram	7
BS-2	2	Disassembly Methods	12
BS-3	3	Cleaning and Lubrication Methods	22
HT-1	4	Amplifier Schematic Diagram	29
HT-2	5	Receiver Schematic Diagram	41
HT-3	6	Installation Methods	52
LL-1	7	Installation Diagram	60
LL-2	8	Wiring Diagram	63

This system has two other advantages over waiting until the last minute to sort out illustrations: each contributor to the document can make references to the figures and tables within the text according to the preliminary numbers, and using the "search" feature that nearly all word processors include, the document coordinator can automatically change the preliminary numbers to the final numbers in the production copy.

Rewriting

Rewriting, the stage in which the real work takes place, should be done in six stages:

- Determine an overall structural pattern while reorganizing the piece as required and checking for continuity.
- Check for rhetorical integrity, rewriting as necessary.
- Eliminate logical fallacies.
- Recast awkward sentences, while inserting transitional words and while paragraphing for plenty of white space.
- Remove spelling and grammatical errors, and correct the punctuation as necessary.
- Copyedit the entire document if coordinating a group effort. If not, perform a final check.

Determine an Overall Structural Pattern

You should begin this stage of rewriting by skimming rather than by reading carefully. Look for logical gaps and for breaks in the smooth flow of thoughts. Part of the job of producing a second draft is to decide what large chunks (at least paragraph size) should be moved from one place to another, what parts should be deleted, and what parts should be added. You may well find that you will have to revise your outline at this point. In consultation with your co-workers, rewrite the outline as necessary. Once again, don't become a slave to an outline that doesn't work. Many word processing programs include an outlining feature that you might find helpful.

Check for Rhetorical Integrity

The seven liberal arts of a classical education included exhaustive study of the *trivium*, which is composed of grammar, rhetoric, and logic. The non-classical approach to education, particularly over the past 60 years, has tended to emphasize rhetoric at the expense of logic and grammar. However, the ethically composed technical document provides proof that a balance can still be struck.

Engineers and technicians deal in the hard currency of bridges that the public expects will stay up in a stiff breeze, of spacecraft that astronauts confidently mount in the hope they will orbit, and of computers that users suppose will add up figures correctly. The occasional failure serves only to throw into sharp relief the importance of accurate work and ethical conduct within the technical professions. Consequently, straightforward, logical, ethical, and balanced technical documents are not a luxury. They are an absolute necessity. What do you need to do in the rewriting stage to ensure that your document exhibits these qualities?

Rhetoric is sometimes called the art of persuasion. One practices an art by discovering its principles and by applying them intuitively. The principles of rhetoric are few:

1. In general, choose words and word combinations to impart the tone and meaning you intend.
2. Use the position of natural emphasis to help convey thoughts.
3. Choose powerful verbs to move readers forward.
4. Use reader-centered methods of exposition, including graphic illustration to impart spatial information, to minimize the expenditure of reader energy.
5. Use tables to impart most parallel information, especially to compare data.

CHOOSING WORDS AND WORD COMBINATIONS

The words you choose are not like numbers or metric symbols. They make sounds that contribute to the reader's overall impression and response to the message, and they include a quality we refer to as connotation. When we say that a piece of writing has a certain tone, we refer to the sounds the words make and the connotations they transmit.

For example, let us suppose you begin a letter in response to a customer's complaint with the following sentence:

> We are in receipt of your letter in which you claim the paint on your Gasguzzler V8 has faded, and we will respond in due course.

You may think you are merely stating facts. Your customer, on the other hand, may look upon the same statement as an accusation. "What does that clown mean by *claim?* It *has* faded! *Due course?* They're going to cook up some excuse to avoid paying for a new paint job!"

Be aware of the connotations of the words and structures you use. And especially watch for words or structures that further a *bandwagon* response or a *false dilemma* frame of reference (see Chapter 5).

USING NATURAL EMPHASIS

The position of natural emphasis falls at the end of each structural unit: that is, the end of the sentence, the last sentence in a paragraph, the last paragraph in a section, and the last section in a document.

Natural emphasis affects all technical documents, some more than others. It is less important in manuals and other instructional writing than it is in reports and proposals. Therefore, when you are rewriting reports or proposals, you will find that repositioning key information or points of particular significance can often add punch to the message you wish to convey. Consider the following example:

> The performance of the production unit differs from that of the prototype because the turbine blades have been redesigned. The extra heavy rear spar is not included in the production model, since the resonant frequencies are shifted into another range by the new blade design.

Although the example is clearly written, the writer has not paid attention to filling the positions of natural emphasis with the significant parts of the message. Is the improved performance significant? Is the absence of the extra heavy rear spar an advantage? If so, the writer might have rewritten this paragraph as follows:

> Redesigned turbine blades have improved the production unit's performance beyond that of the prototype. Furthermore, since the resonant fre-

quencies of the redesigned blades lie outside the range of other structural components, we have been able to eliminate the extra heavy rear spar.

Rewriting sentences and paragraphs in this way is painstaking work, but you will find that the effort required will impart a professional polish to your document.

CHOOSING POWERFUL VERBS

Powerful verbs move the reader along by conveying meaning in concrete terms. Compare the following two versions of the same paragraph:

> The universe is full of things that take care of themselves. Atoms spontaneously assemble into molecules. Mountains steadily rise out of the sea. Stars continuously cluster into galaxies. Yet nowhere is self-organization more impressive than in crystals.[1]

> The universe is full of things that take care of themselves. Molecules consist of atoms. Mountains come into being spontaneously. Galaxies are composed of stars. Yet nowhere is self-organization more impressive than in crystals.[2]

Notice that although the second, paraphrased version is shorter, it does not communicate the idea nearly as well as the original, which contains the verbs *assemble*, *rise*, and *cluster*. These verbs convey images, whereas the substitutes *consist of*, *come into being*, and *are composed of* are weak and abstract. And please note that none of the substitutes is a passive construction. Some people think that the passive construction is the greatest enemy of good technical writing. They are wrong. The passive construction, which can be very useful if not overused, is only one of the weak verb formations that can drain the life from technical documents.

USING READER-CENTERED METHODS[3]

Example

Examples, most useful for bringing theoretical discussions into focus, can make important facts more easily understood than they would otherwise be because they relate them in a sensible context.

---

[1] Fagan, Paul J. and Michael D. Ward. *Scientific American*, July 1992, p. 48.

[2] Paraphrased version of the same paragraph with three verb changes.

[3] Refer also to Chapter 5.

Comparison

Comparisons may be made in several ways. Tables are extremely useful for comparing data, as are several kinds of illustrations. The focus in this section, however, is on comparisons as an integral part of text. Tables are discussed later under a separate heading.

Analogy

Analogy is a kind of comparison, except that the relationship between the two things compared is far apart. The value of the analogy comes about because of the reader's familiarity with the item chosen for comparison or because of the ease with which the reader can picture the item. Analogy is very useful in representing abstract or theoretical processes.

Graphic Illustration

Of all the devices available to the technical writer, graphic illustration is the most valuable. In fact, it is indispensable. (An illustration is any line representation, photograph, diagram, or chart.)

A technical writer should never forget, however, that the text must always bear the burden of continuity. Tables or illustrations that are merely dumped into the text force the reader to shift without warning from the *linear* arrangement of text to *parallel* or *spatial* presentations. Tables are normally arranged in parallel fashion, and illustrations are spatial by nature. The only reader-centered way to integrate these non-linear presentations is to refer to them in the text *before* including them in the document. In this way, the overall linear character of the document is maintained. Therefore, good technical writers always refer to tables or illustrations before placing them in the text.

Having referred to an illustration or table, the good technical writer then places the table or illustration in the first available space following its reference. If space permits, partial-page illustrations should appear on the page containing the reference. If insufficient space remains following the reference, the illustration should be placed at the top of the succeeding page.

An illustration should always include a figure number and title within its caption, and the caption should appear immediately beneath the illustration.

A full-page illustration belongs on the page following its reference. If a writer must make reference on the same page to more than one full-page illustration, the resulting pile-up of illustration pages can interfere with the smooth flow of text. For this reason, partial-page illustrations are better than full-page illustrations.

USING TABLES TO IMPART PARALLEL INFORMATION

Tables enable the writer to place comparative data side by side. Consider the relative merits of the following pieces of writing, the first in prose alone and the second with an accompanying table.

> Three copier brands were considered for our office: the Super 101, the Huxley 28, and the Darwin 12. The Super 101 is the most expensive of the three machines at $6500, but its rate of 76 copies per minute is best. The Huxley 28 will duplicate only 27 copies per minute, while the Darwin 12 will produce 46 copies in one minute. The Huxley is the cheapest of the three at $1200. The Darwin is $3400. Another advantage of the Super 101 is the warranty, which is for a full 12 months. Both the Huxley and the Darwin warranties are for only 90 days from date of purchase. The Darwin dealer offers a first-year service contract for an additional $125. The Huxley dealer does not offer any service contract agreement.
>
> For our purposes, the Darwin 12 appears to be a good choice because of its reasonable copying rate and its modest cost.

And now, here is the version with an accompanying table. Note that the table reference appears before the table.

> Three copier brands were considered for our office: the Super 101, the Huxley 28, and the Darwin 12. As shown in Table 1, the Super 101 is the costliest and fastest of the three machines, while the Huxley 28 is the cheapest and slowest. Since the Darwin 12's speed of 46 copies per minute would be adequate for our use and since the dealer offers a service contract at modest cost, this machine appears to be the best value on a cost-per-copy basis.

TABLE 1   COMPARISON OF OFFICE COPIERS

Machine	Warranty	Cost ($)	Copies/Min.
SUPER 101	12 months	6500	76
HUXLEY 28	90 days	1200	27
DARWIN 12*	90 days	3400	46

* Additional 12 months for $125.

Eliminate Logical Fallacies

LOGIC IN A TECHNICAL DOCUMENT

A technical document is logical if the conclusions follow from the data presented. As a technical writer, however, your ethical responsibilities range beyond the document. Most of us are familiar with the so-called "docudrama." These television programs are prime examples of structures designed to support conclusions that follow inevitably from the information presented. Their purpose, however, is not to impart knowledge but to act as a commodity in the market for sensational programming. Technical writers who select only the data that supports their conclusions present a biased view also. The temptation to carry on this type of unethical practice looms large, especially in proposals and in certain kinds of reports.

Bearing in mind the ethical responsibilities of a technical writer (discussed fully in Chapter 16), you must root out the fallacious reasoning that might result from a careless style.

Recast Awkward Sentences and Paragraphs

The beauty of working with a word processor is that changes are immediate and the cluttered draft that can create confusion is eliminated. Still, some people like working from a printed draft. If you are one of those people and if the changes thus far required are extensive, you may wish to produce a clean draft before going on with this step. If the copy is fairly clean, you can complete this step before producing the second draft.

Recast sentences that are too long, perhaps breaking them into two or more sentences. Insert transitional expressions, such as, *however, therefore, nevertheless,* and *consequently,* wherever the flow of the writing will be improved by them. For example, if a causal connection exists between two paragraphs, the word *therefore* at the beginning of the second paragraph will cue the readers to the causal relationship, thereby saving them a mental step.

Paragraphs should not be so long that they intimidate readers. Short paragraphs make for easy reading, whereas long paragraphs often appear tedious. Therefore, break long paragraphs into two or more smaller paragraphs. Logical break points can nearly always be found to justify short paragraphs.

Remove Spelling and Grammatical Errors

SPELLING

English spelling is difficult for some people, and your first efforts to find spelling errors may be disheartening if you are one of these people. Most people can learn to spell, but it takes time and effort. Software spell checkers are useful, but most of them are not foolproof. For example, if you write *affect* when you mean *effect,* the less sophisticated spell checkers will not recognize the error. Therefore, you should use these aids to spelling but not rely on them completely.

GRAMMAR

How Important Is Grammar?

Is poor grammar a distraction? Is it an indication that the writer is not particularly literate? Reference Software International of San Francisco hired an independent research firm to find out. In their poll of 100 executives of *Fortune 1000* corporations, the researchers compiled the following information:[4]

1. Eighty percent said that they have, on occasion, decided against interviewing job candidates, based solely on poor grammar, spelling, or punctuation in the application letter or résumé.

2. Sixty-three percent feel that poor writing skills reflect poorly on the writer's intellectual ability.

3. Ninety-nine percent agreed that poor writing and grammar are a hindrance to a person's chances for promotion.

Although poor grammar distracts readers, reflects negatively on the writer or speaker, and indicates the writer's lack of knowledge or intellectual capacity, its real significance lies even beyond these troubles. Grammatical correctness is not, as some people believe, merely a cosmetic addition to a communication. The hard truth is that *grammar and content are inseparable.* Poorly structured technical documents have produced everything from unreliable toasters to launch-pad explosions.

Although thinkers have known for several centuries about the inseparable nature of the content of a piece of writing and the verbal structures that reveal it, many people still make the error of separating them. Those people believe that grammatical structure is a cosmetic addition, and they sometimes talk about the "value added" nature of a technical writer's work, an idea implying that a piece of writing has value (content) before it has structure.

Applied to sculpture, this same reasoning would lead us to conclude that a block of marble contains a statue before the sculptor touches it. You may have heard the old joke about a sculptor who, when asked how he was able to produce a figure of a horse that was amazingly lifelike, said that he just began with a block of marble and chipped away everything that did not look like a horse. This joke is funny (not very) because the idea that the horse is in the

---

[4] *Training,* August 1992.

marble in the first place is ludicrous. The sculptor cannot make a marble sculpture without marble, of course, but marble's essential character is not the same as that of a marble sculpture of a horse.

Like the sculptor's block of marble, a technical expert's draft that fails to communicate unambiguous information has neither content nor value beyond that which the technical writer may eventually give it. Therefore, we may conclude that while the technical editor may *add* value, the technical writer *creates* it.

The notion that structure and content are separate is widespread. Those who hold the idea say that, like a technically imperfect musical performance with power and passion, a technically imperfect piece of writing may move us deeply. That argument has currency in one sense, but it is profoundly misleading in another. Only the *emotional* content of a piece of writing is inextricably bound up with the character of the writer (or fictional character in the case of fiction) and can thus be conveyed by "imperfections" of grammar and style. A similar quality in technical writing will only obscure the intended message — if the author's thoughts are clear enough to formulate a message in the first place. When practice has brought you to a full appreciation of the distinctions within writing style, you will see that only parts of style and content are separable.

Finally, imperfection in a technical document is also completely different from the "imperfection" of a musical performance. For all its subtlety, the playing technique of a musician is made up of discrete sub-techniques that have a characteristic sameness because they all relate to physical movement. The "music" comes from the musician, not the musician's technique, which releases the music without creating it.

Will a Grammar Checker Help?

Analysis for grammatical structure is an essential part of the rewriting process. Again, this step is not a matter of cosmetic arrangement; in fact, it is a "rethinking" process amounting to an analysis of the message itself. Grammatical structures parallel logical ones and, in so doing, ultimately create the message. Computerized grammar checkers are no substitute for a knowledge of grammar. In fact, they merely confuse and confound those who don't understand grammar and waste the time of those who do. The following example illustrates the shortcomings of computerized grammar checkers:

Is Grammar Checkers Useful?

Grammar checkers in general is an abomination and a snare. There is two good reasons for avoiding them entirely: their inconsistent flagging of agreement errors and their inherent inability to correctly identify subjects.

Run the title of this paragraph through your grammar checker. Mine, supplied with my Microsoft Word application, didn't find anything wrong with it. An author who doesn't have a good understanding of subject-verb agreement rules might let that major error slip by. Neither did the grammar checker find anything *grammatically* wrong with the two sentences making up the paragraph. Having failed to pick up the agreement error in the first sentence, however, my grammar checker went on to challenge (incorrectly) the expression "in general," which its maker considers a stylistic miscue. It also failed to note that the second thought is cast in an expletive sentence construction, and it treated the adverb *there* as a subject. Why would a grammar checker not "know" that an adverb cannot act as a subject under any circumstances? Obviously, its maker based the program on outdated and severely flawed notions of grammar.

Here's what the same grammar checker does with a typical sentence containing a compound verb.

> She has served as a president, secretary, and treasurer and has distinguished herself in all these posts.

Because the grammar checker once again failed to identify the subject, it suggested the word "have" for "has" in the second half of the compound verb structure, thereby encouraging the user to *introduce* a grammatical error.

Many, maybe all, grammar checkers consistently ignore grammar while citing stylistic choices that poorly educated people identify as grammar errors. A careful look at how the grammar checker seems to be going about its job leads one to the realization that the fault lies not with the program designer, but rather with the "authority" that was consulted in an effort to find out what a grammar checker should do.

Do You Have a Developed Grammatical Sense?

The heading to this paragraph is a trick question. If you can speak and understand speech, you definitely do have a grammatical sense. But you may well have to develop that sense and look objectively at what you know and don't know before you are able to write skillfully.

Even if your grammatical sense is above average, your may not recognize serious grammatical errors at once. If you are rewriting your own draft and if time constraints are not critical, wait several days before rereading it. You will see your work with fresh vision after a few days. Even then, however, merely reading the document to try to pick up grammatical errors will prove to be an unproductive activity unless you possess a finely developed grammatical sense.

How Can You Develop Your Grammatical Sense?

The best way to grammatical competence is to approach this stage of the rewriting process with a set of criteria in mind. Just as a doctor must diagnose an illness before prescribing a remedy, you must diagnose the grammatical trouble before changing words and sentences. Those who rely on their ears to tell them when something is wrong will, more often than not, confuse grammar and style. For these reasons, you should become familiar with the eight specific types of sentence errors listed below and fully described in *A Self-Directed Course in English Grammar and Usage*, Section V of this book.

ELIMINATE DANGLING AND MISPLACED MODIFIERS

Have you used any locutions similar to the following?

> Using a digital approach, it is possible to monitor input errors to all entry points in the system.

> Shifting from ice to hydrated minerals, the general conclusion is the same.

> Using existing methods, the roller detent assembly will be assembled onto the thrust bearing housing.

The foregoing examples all contain *dangling modifiers*. Logical consistency demands that a modifying phrase at the beginning of a sentence refer to the grammatical subject, and the subject of each of these sentences is unsuitable for such a modifier. In the first example, the expletive "it" has no antecedent; therefore, the modifier "using a digital approach" doesn't refer to anything.

In the second example, "the general conclusion" isn't "shifting from ice to hydrated minerals."

And the third example is a passive construction beginning with a participial phrase. If you consider that the subject in a passive sentence receives the action and that the real actor in a passive construction is contained in a *by* phrase (whether the phrase is present or implied), you can see that this example should read as follows:

> The roller detent mechanism will be assembled onto the thrust bearing housing by technicians using existing methods.

Or, if you would rather use an active construction, you could write the following:

> Using existing methods, technicians will assemble the roller detent mechanism onto the thrust bearing housing.

Or, if the intention is to write a procedural step, you would write the step in the second-person, imperative mood:

> Using existing methods, assemble the roller detent mechanism onto the thrust bearing housing.

In the example above, the participial phrase modifies the understood subject *you*.

A *misplaced modifier* is one that is not close enough to the word it modifies. The following example illustrates the problem:

> Heat treating of the forged casting once was considered to be a possible cause of the crankshaft failure.

Does the modifier *once* refer to *heat treating*? Or, more likely, does it refer to *was considered*? The problem may be solved in several ways:

> Heat treating of the forged casting was once considered to be a possible cause of the crankshaft failure.

> At one time, heat treating of the forged casting was considered to be a possible cause of the crankshaft failure.

MAKE LOGICALLY EQUAL STRUCTURES GRAMMATICALLY EQUAL

Aristotle discovered the subject of logic some 2500 years ago. He was not the first person to think logically; he was just the first to study the ways that thoughts go together. But he could not have made the discovery if human beings were not logical to begin with.

This logical characteristic of human beings leads us naturally to prefer reason over arbitrariness, harmony over cacophony, and order over chaos. We feel abused when we are subjected to an unjust law; we feel uncomfortable when we are exposed to discordant sounds; and we are distracted when we are presented with grammatical structure that does not fit the logic of the intended message. The distraction caused by non-parallel grammatical structure has nothing to do with our knowledge (or lack thereof) of grammar. If someone says, "I like to play tennis, golf, and boating," you don't have to be a grammarian to realize that this person doesn't really like, " ... to play ... boating."

So what harm is there in such a construction? We can certainly know what the sentence means. Why should we nitpick? Isn't non-parallel construction nothing more than a minor annoyance? The answer is that *all non-parallel constructions reduce the probability that communication will occur by making the reader do an unfair share of the work* and that *some non-parallel constructions create a false message.*

Most people respond to annoyance by avoiding it. Why should a reader be made to do an unfair share of the work of communication? Most readers unconsciously assume that a lazy writer doesn't deserve to be read. And they are right. A writer who doesn't carefully structure the message is saying by that carelessness: "This message isn't important." Why should a reader oppose such a clear statement?

In technical writing, the problem of non-parallel structure can have consequences that go far beyond annoyance. Consider the possible consequences in the following example in which a person/mood shift destroys parallel structure:

1. Open all the access doors on the ejector rack fairing.
2. Both the cartridges should be removed from the dual breech chambers.
3. Remove the ejector rack safety pin.

Here we have three simple procedural steps having to do with removing an external store (bomb, fuel tank, etc.) from an aircraft. The first and third steps are in the second person, imperative mood, but the second step is in the third person, subjunctive mood. In this case, the non-parallel structure can become more than an annoyance or an impediment to efficient communication if the technician following these instructions assumes that removal of the cartridges is optional.

If the ejector rack fires an external fuel tank (or something worse) into the parking ramp while the technician carries out the procedure, the matter will have gone beyond the stage of mere annoyance.

All readers, most of them unconsciously, expect that the logical connections between thoughts and the grammatical structure of those thoughts will match one another. When logic and grammar are mismatched, expectations are thwarted, forcing readers to use extra time and energy to supply the missing connections. Although the time required may be only milliseconds and the energy required may be immeasurable, either is often sufficient to cause misreading. At best, such errors create distractions that drain away the power of the message.

When you are rewriting, pay special attention to procedural steps. Make certain that they are always in the second person, imperative mood. Also, look for words and expressions connected by *and*, *or*, and *but*. These words will tell the reader that the words and expressions thus joined are logically equal. Therefore, you must make certain that they are also grammatically equal. Here is an example:

The money available will permit the redesigning of the unit or a partial development of a replacement unit.

In this example, we have the gerund *redesigning* in parallel with the noun *development*. Such an error is relatively minor in itself, since a gerund is by definition a verbal that takes the place of a noun. However, writing in which such thoughtless phrasing is common becomes difficult to read as a result of the accumulation of small errors.

Elliptical constructions sometimes lend themselves to non-parallel structure.

> The tests were successful and the engineer vindicated.

This sentence asks the reader to believe that "... the engineer *were* vindicated." The sentence should read as follows:

> The tests were successful, and the engineer was vindicated.

Although most elliptical constructions will not lead to errors of this kind, a good rule-of-thumb for technical writers is to include all the words and to leave elliptical constructions for newspaper editors, who may need such shortcuts to save column space.

ELIMINATE UNNECESSARY SHIFTS IN PERSON, MOOD, AND VOICE

Technical writing is most often in the third person, indicative mood, although instructional writing is sometimes in the second person, imperative mood. (Please refer to Section V, A Self-Directed Course in English Grammar and Usage, for an explanation of grammatical terms with which you are unfamiliar.)

This textbook is an example of instructional writing. Therefore, shifts from the third person, indicative, to the second person, imperative, are common throughout. The steps contained in instructional manuals and process sheets should always be in the second person, imperative. An example within the preceding discussion on parallel structure demonstrates how a shift away from that form can cause trouble.

Proposals and reports are examples of non-instructional writing. Therefore, shifts from the third person or from the indicative mood should not be made without good reason. If the document you are rewriting is either a proposal or a report, examine each paragraph carefully to make sure that such shifts are absent. The following example is from a fictional proposal:

> The Bathythermograph System will comprise three major subassemblies: the airborne launch system, the expendable buoy, and the processing equipment. The airborne launch system will be installed as an integral part of the aircraft, and maintenance procedures will be carried out as part of the 100-hour inspection. The expendable buoy will not require maintenance; it will be factory inspected, and its "use-before" date will be the only criterion for ascertaining its operational status. However, make certain that the processing equipment is operational before a mission is undertaken.

The final sentence in the example is out of character with the rest of the paragraph, forcing readers to shift their point of view. Will the reader actually have to perform this operational check? The sentence should have read: "However, the operator will have to ...."

Shifts in voice occurring within sentences are particularly distracting because they require the reader to see two points of view in a single sentence.

Here is an example:

> The information must be stored immediately, but operators must complete the run before removing the disk.

Readers are made to shift from a passive-voice construction in the first clause to an active-voice construction in the second clause. In fact, readers are made to do more than just mentally shift gears. They must wonder also whether the operator is responsible for storing the information. The sentence does not say so. The problem may be solved in several ways:

> Operators must store the information immediately, but they must complete the run before removing the disk.

> Although operators must store the information immediately, they must complete the run before removing the disk.

The first rewritten version places equal weight on both actions. The second relegates the storing action to a dependent clause, thereby indicating it to be somewhat less important than the completing action. If the completing action is the more important of the two, this second version is the better statement.

USE SUBORDINATION TO CLARIFY CLAUSE RELATIONSHIPS

The final version of the foregoing example shows how a subordinate clause can enhance meaning. Our reader not only expects that we will put equal thoughts in equal grammatical structures, but that we will not put unequal thoughts in such structures. Therefore, look for non-parallel ideas contained in parallel grammatical structures. Here is an example, with the coordinated version first and some possible subordinated versions in parentheses:

> The loads reach equilibrium, and the mechanism shuts off.

> (When the loads reach equilibrium, the mechanism shuts off.)

> (The mechanism shuts off when the loads reach equilibrium.)

> (If the loads reach equilibrium, the mechanism shuts off.)

> (The mechanism shuts off if the loads reach equilibrium.)

Any of the foregoing versions may be desirable. The emphasis and tone intended would dictate the choice.

USE ACTION VERBS AND ACTIVE VOICE WHENEVER POSSIBLE

Look for any tendency to overuse the verb *to be;* however, don't overcompensate. There is nothing inherently wrong with linking verbs, but the more action verbs you use, the more animated your writing will be. If you spot any paragraphs that contain only linking verb constructions, try to change at least a few of them.

Also, look for overuse of passive-voice constructions. The need to be "objective" does not necessarily imply the passive voice. Here are two examples, again with rewritten versions in parentheses:

The tests were failed by six of the units in the first batch.

(Six of the units in the first batch failed the tests.)

For evaluation purposes, five preproduction models were constructed.

(The technician constructed five preproduction models for evaluation.)

Using the active voice provides a secondary benefit, that of reducing the total number of words.

USE COMMAS CORRECTLY

Correcting the punctuation to properly place commas will be easy if you subscribe to a few simple rules. (Refer to Section V, A Self-Directed Course in English Grammar and Usage.) Unfortunately, many writers have accepted the gradual erosion of comma rules. In fact, some writers, including "professional" writers, think that there is something inherently wrong with commas. Here is what Gertrude Stein had to say about commas:

A comma by helping you along holding your coat for you and putting on your shoes keeps you from living your life as actively as you should lead it and to me for many years and I still do feel that way about it ...[5]

This sort of playful approach to language is great if you are Gertrude Stein. If you are concerned about sound technical writing, you will wish that Gertrude Stein had taken up a different line of work.

At least two reasons can be found for the confusion about comma usage. The first is the mistaken belief that a comma represents the smallest interruption in speech. In fact, *the comma represents the smallest interruption in continuity of thought, and thought patterns seldom coincide with speech patterns.* If you have been taught to place a comma where you would pause in speaking the words you have written, erase that notion from your memory before referring to the rules given in Section V. If you don't, you will never learn to use commas correctly.

The second reason for the aversion to commas is perhaps the other side of the same coin. Some writers try to avoid commas, believing that commas prevent smooth reading. This belief rests on the idea that a comma signals readers to pause in their reading. However, you will recognize the error in this thinking by reflecting on the comma's purpose, which is to separate thoughts and not to establish speech patterns.

---

[5] Stein, Gertrude. *Lectures in America*. Beacon Press, 1957, p. 219.

Commas do not inhibit good readers; on the contrary, the presence of correctly placed punctuation enables all readers, good and poor, to read much faster than they could otherwise. Even when reading comma-free prose that has been carefully constructed to avoid ambiguity, readers must expend extra energy if they are to see logical separations based on word juxtaposition alone. And in poorly constructed prose, more than just added energy is required. Sometimes only mental telepathy will do the trick!

The irony in the liberalization of comma rules is that the less power the rules have, the more difficult writing becomes for those people who still care about communicating well. The plain fact is that the rules, and not the liberalizers of rules, make the writing job easier. So learn the rules, and rejoice in your mastery of them. They will help you to communicate clearly, quickly, and efficiently.

Copyedit the Entire Document

If the writers edit their own drafts, the benefit to be derived from the final reading will be directly proportional to the amount of time that has elapsed between the writing and the editing. For this reason, a reading by someone who has yet to see the copy is often the most productive, provided of course that the person chosen is a good reader and is skilled in English structure. When you are a coordinator of a group, you will either perform this task yourself or designate someone in the group as a copyeditor.

Changing someone else's writing is a matter that requires knowledge and consideration. For this reason, copyediting demands a disciplined approach. The copyeditor who cannot distinguish between structure and style is the bane of all authors, technical or not. Copyeditors who cannot "hear" the author's voice and who have no sense of rhythm can nonetheless perform a valuable service by catching spelling and minor grammatical errors, provided that they have a good knowledge of spelling and grammar. But the copyeditor who cannot tell you *why* your copy needs a certain change will often render great damage to your manuscript. These editors rely exclusively on their "ears" to tell them whether something needs changing. But thus used, the untuned ear is a deadly weapon. Copyeditors who fancy themselves to be stylists will create one of three bad results: change the message, diminish its power, or alienate the authors while adding to their workload.

Good copyeditors restrict themselves to the flagging of faulty grammatical structures. Furthermore, they know precisely what fault is exhibited and the most economical way of eliminating it.

But unskilled copyeditors may flag all sorts of unnecessary passages. They might, for example, flag all split infinitives because they do not know that infinitive phrases in a largely non-inflected language such as English very often *require* splitting for correct modifier placement; they may even sometimes introduce ambiguity by "correcting" this perceived fault. Jacques Barzun summarizes the problem very nicely — if a bit angrily:

> *Thus split infinitives (or what are wrongly thought to be infinitives), sentences ending with a preposition, paragraphs beginning with* And *or* But *. . . are objects of copy-editor persecution.*[6]

Do not fall into any of the foregoing traps. Instead, recognize that your responsibilities as a copyeditor include a thorough knowledge of the sentence faults mentioned above and clarified further in Section V, A Self-Directed Course in English Grammar and Usage.

A Trip Through the Process

Let's look at a typical job at each stage in the process. Let us suppose that you work for an automobile manufacturer and that your boss, Division Manager Ethel Billingsgate, tells you to investigate the increase in absenteeism and the decrease in an already low productivity at the facility that builds torque converters.

During your visit to the torque converter plant, you discover that the windows are all painted black. When you ask the plant superintendent why the windows are painted, he tells you that people don't need to be gawking out the window when they are supposed to be working. He also volunteers that they are all a bunch of slugs and that if they aren't constantly hounded, they don't work.

You find out also that, although most of the operations in the plant are automated, the final assembly area, where most of the people work, is right next to the press room; you note that the noise is deafening in this area. You also observe that the tool crib is at the far north end of the building and that, at least half the time, three or four machine operators are waiting in line there. You calculate the number of worker-hours that this method of dispensing tools is costing the company in lost labor.

You then enlist the aid of the shop steward and distribute a questionnaire that asks employees what working conditions they think need changing. Most of the responses cite the poor lighting, noise, and high dust concentrations in the plant.

---

[6] Barzun, Jacques. *A Word or Two Before You Go.* Wesleyan University Press, 1986, p. 86.

Back at your desk in your airconditioned office in the engineering building, you telephone the millwright superintendent for an estimate of the cost of moving the final assembly line, installing an air filtration system, and putting in fluorescent lights. Since the report will be a short memorandum, you decide on a simple problem-solution oganizational pattern. Exhibit 6-1 shows the first draft of the memorandum report you will eventually send to your boss.

First Draft of a Memorandum Report

MEMORANDUM

To: Ethel Billingsgate   **Date:** March 5, 1997
     Division Manager

From: Thelma Tryhard   **Subject:** Torque Converter Plant

I found a number of problems while following up your request that I look into the absenteeism and poor production record at the torque converter plant. The plant superintendent seems to be a pretty heartless person, having painted the windows so the workers can't look outside thus raising the absentee rate by 20 percent. The workers have a very low morale because of the bad relationship with this very authoritarian superintendent. We need a new Plant Superintendent. The workers themselves are unhappy with the poor lighting and dust in the air. Also, the tool crib arrangement and the assembly line next to the press room are a source of trouble. If we had the paint taken off the windows, the absentee rate would be less, but I'm not sure the lighting would be adequate anyway, since the incandescent lighting system doesn't offer much illumination. Maybe we need to have a fluorescent system to solve the lighting problem. We also need the tool crib to be closer to the machines whose operators use it. The problem with the air could be corrected if we were to install an electrostatic filter. And the plant needs to be reconfigured to get the final assembly line away from the noise of the press room. All of this would cost about 350,000 dollars, but it should be worth it in the long run.

Looking at the first draft, you realize that it lacks any structure at all and that it does not even exhibit the top-down organization a good report would have. In fact, even the subject line needs work. After *reorganizing and determining the structural pattern,* you decide to put in some headings and arrange your paragraphs under them. Then you examine the piece for *rhetorical integrity and logical fallacies* and come up with a second draft that includes some better word choices, stronger verbs, significant information in positions of natural emphasis, and some reader-centered methods of exposition. Exhibit 6-2 shows the result.

EXHIBIT 6-2	Second Draft of a Memorandum Report

MEMORANDUM

To: Ethel Billingsgate
     Division Manager

Date: March 5, 1997

From: Thelma Tryhard

Subject: Causes of Absenteeism and Low Productivity in the Torque Converter Plant

ABSTRACT

The problems in the Torque Converter Plant result from poor working conditions. The solutions require changes to the physical plant and familiarizing the plant superintendent with modern management techniques.

CAUSES OF ABSENTEEISM

The plant superintendent has a poor relationship with the employees. He is an authoritarian with an approach to management that suggested to him that painting the windows so the workers can't look outside was a good idea. The absentee rate went up by nearly 20 percent after the windows were painted, indicating a possible connection between the poor lighting and absenteeism. An excessive amount of dust in the air may be exacerbating respiratory ailments and thus adding to the absentee rate as well.

CAUSES OF LOW PRODUCTIVITY

In addition to absenteeism the plant layout does not lend itself to efficient operation, which is a major cause of low productivity. The final assembly area, in which most of the employees work, is located next to the press room. The noise from the presses makes concentration nearly impossible. Assembly errors occur more often than statistics would predict, suggesting that the excessive noise adversely affects the operation of this department. Also, the tool crib is located at the far north end of the building, and this location requires a lot of travel time for the machine operators.

continued →

RECOMMENDED SOLUTIONS

During the next model changeover, while the plant is shut down for several weeks, relocation of the assembly line away from the press room would be a good idea. At that same time, we should remove the paint from the windows, install fluorescent lighting, and put in an air filtration system. The plant superintendent should be sent to the next series of management seminars. We could hire one more tool crib employee for each shift, who could operate a tool delivery operation, thus saving us 15,000 worker-hours per year. The plant modifications would cost $350,000, but would lead to a potential saving of $180,000 per year. The salaries of the three additional tool crib employees would be approximately half of the worker-hours currently lost through our present method of dispensing machine tools.

The second draft represents a big improvement. We now have something that will lend itself to refinement into a first-class technical document. Our next steps are to recast awkward sentences, insert transitional words and phrases, and paragraph the piece for appearance. Looking at the abstract, we see a non-parallel structure and poor subordination. Next, we find that we have used more words than necessary in the first paragraph and that it lacks proper subordination. The second paragraph needs work too; for example, it begins with a sentence that contains a dangling adjective clause. Our "solutions" paragraph exhibits several bad modifier placements, as well.

Having completed those modifications, we can check for spelling, remove any grammatical errors that still remain, and correct the punctuation. Our final draft might then look like the one that follows:

EXHIBIT 6-3 **Final Draft**

MEMORANDUM

To: Ethel Billingsgate
    Division Manager

From: Thelma Tryhard

Date: March 5, 1997

Subject: Causes of Absenteeism and Low Productivity in the Torque Converter Plant

continued ➜

ABSTRACT

Absenteeism and low productivity in the Torque Converter Plant result from the Plant Superintendent's outdated management philosophy, which has led, in turn, to poor working conditions. The solutions require changes in management style and in the physical plant.

CAUSES OF ABSENTEEISM

The Plant Superintendent has a poor relationship with his employees. His authoritarian approach to management even led him to paint the windows so the workers can't look outside. The absentee rate went up by nearly 20 percent after the windows were painted, a rise that may indicate poor lighting as a cause of absenteeism. An excessive amount of dust in the air probably adds to respiratory ailments and thus to the absentee rate as well.

CAUSES OF LOW PRODUCTIVITY

In addition to absenteeism, which is a major cause of low productivity, the plant layout does not lend itself to efficient operation. The final assembly area, in which most of the employees work, is located next to the press room. (See Figure 1.)

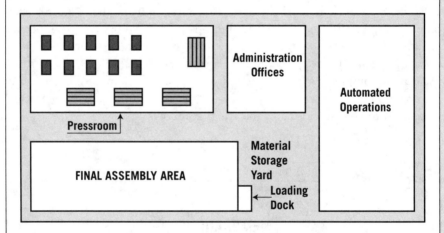

FIGURE 1 — CURRENT SETUP WITH FINAL ASSEMBLY ADJACENT TO PRESS ROOM

    The noise from the presses makes concentration nearly impossible. Assembly errors occur more often than statistics would predict, suggesting that the excessive noise adversely affects the operation of this department. We could move the final assembly area to the location shown in Figure 2.

continued →

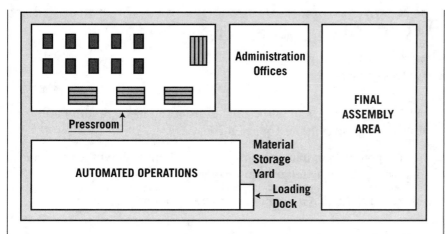

FIGURE 2 — PROPOSED RELOCATION OF FINAL ASSEMBLY AREA

Also, the location of the tool crib at the far north end of the building and our current method of dispensing tools wastes approximately 6000 hours of machine operators' time annually.

RECOMMENDED SOLUTIONS

I recommend that we take the following steps during the next shutdown for model changeover:

1. Relocate the assembly area away from the press room.
2. Remove the paint from the windows.
3. Install fluorescent lighting.
4. Put in an air filtration system.
5. Send the Plant Superintendent to the next series of management seminars.
6. Hire one more tool crib employee for each shift for the purpose of instituting a tool delivery operation.

Although the recommended plant modifications would cost $350,000, they would save the approximately $180,000 per year currently lost through absenteeism. The salaries of the three additional tool crib employees would be $93,000 annually, less than half of the amount our present method of dispensing machine tools wastes in worker-hours.

I can supply you with the written estimates from the Millwright Department for the plant changes if you decide to go ahead with my recommendations. I would point out also that the next management series starts on April 15. Please let me know if you would like me to follow up on any of the recommendations.

Exercises

6-1. Copyediting

1. As a copyeditor, examine the final draft of the memorandum report to Ethel Billingsgate in the Tour of the Process.

2. Go through the complete "rewrite" step and think about the recommendations as they may relate to priority, cost, and timing.

3. Make the changes required to incorporate your ideas.

6-2. Following the Steps in the Writing Process

1. Write a memorandum report that tells what should be done to improve your school's cafeteria, sports arena or stadium, registration procedure, or anything else that concerns you.

2. Make at least three drafts, going through all the steps in the writing process.

CHAPTER 7

Production Considerations

OBJECTIVES: This chapter will familiarize you with the steps necessary for getting a document to the printed stage and with the processes available for reproducing your written work. You will also learn about specifications and style guides, which you may be required to follow in writing technical documents on the job.

General Principles

While it is true that a creative instinct is an asset to the technical writer, preparation of good technical documents and speeches tends to be a fairly technical operation. In fact, a properly executed technical document is comparable in some respects to a well-designed bridge or an intricate computer program. Like the engineer and the programmer, the successful technical writer constantly monitors the document preparation process and attends to a myriad of details.

Fortunately, this part of the job can be easily learned. One who wishes to acquire a proper attitude need only learn the basics of the technical writing process, described in the previous chapter, and a bit about graphic reproduction. With this knowledge at hand, one can form work habits that will lead to the sound practices necessary for the creation of good technical communications.

In a 1993 paper, Dr. Elizabeth M. Lynn cited a growing gap between the communication abilities of college graduates and the diverse communication expectations of business.[1] The results of her survey of managers who hire science graduates show the top-rated communication skills:

1. To express thoughts simply and clearly in writing

2. To concentrate

3. To use correct words, tone, spelling, grammar, and punctuation

In a follow-up survey, Dr. Lynn's findings indicated that business managers rated a fourth qualification as a top priority:

4. To understand the importance of deadlines

The Hollywood director whose motion picture gets finished a year behind schedule and costs millions of dollars more than it was supposed to might still end up getting an Academy Award. The technical communicator who is a day late and a thousand dollars over budget will not enjoy similar adulation, but sudden unemployment is a definite possibility.

The following discussions, along with the sample specification and style guide (see Sample 7-1, pp. 111–116), will provide you with the necessary information to form good work habits and to respect the time constraints of technical document preparation.

---

[1] Elizabeth M. Lynn. "How Colleges Are Failing to Prepare Students for Workplace Communication" (paper delivered at the Conference on College Composition and Communication, San Diego, April 1993).

Graphic Reproduction

Although students write papers, they are seldom required to make more than one (or at most a few) copies. In real work situations, multiple copies are the rule, and consideration of the methods by which reproduction will be accomplished is essential.

For example, let's look at a contractual report. The contract may require 15 to 50 or more copies for the customer. A company's internal distribution list may typically specify twice that many, including copies for department heads, contracts administrators, document custodians, field offices, and other divisions. In addition, the National Library in Ottawa will require copies.

The task, then, is not simply to prepare a copy of a document, but to prepare a copy that can be readily reproduced. Accordingly, the first step is to find out what method of reproduction will be used, and the second step is to talk to the person or persons who will be responsible for carrying out the work. If your document is to be printed by the offset method, as some documents are, those persons who will do the actual printing will want to know whether you will be supplying disks, camera-ready copy, or copy that needs typesetting. They will also want to know about any special arrangements that will be needed. Even if your document is published by desktop methods, you will want to know what machines and processes are available.

If the document is to be duplicated by the offset printing method, the person in charge of printing will tell you that more than one color (i.e., black) will add to the time and cost of printing, since color separations will be necessary. You will then realize that your plans for very beautifully drawn graphs in blue, green, and red need rethinking.

Note: Color copiers, which make excellent color reproductions, are available at affordable prices and may be used, of course, if relatively few copies of the document are needed.

Photographs, which necessitate halftone negatives and plates for the offset process, will add to the time and money needed for completion of the document. These added costs may well be justified, but it is the responsibility of the technical writer to know about costs and alternatives and to weigh them against the desired result. Let's suppose, for example, that a 100-page document requires 30 illustrations. The cost of having these drawn could easily exceed the cost of 30 photographs and the halftone negatives and plates required for printing these photographs. The most important consideration, however, is which form would better suit the purpose of the document. When a document is governed by a specification, the choice may not exist at all, since some specifications exclude one or the other method.

Another extremely important aspect of planning a technical writing job is the time necessary for each step in the process. If your document requires offset printing, you may have to allow a week or more for the printer to do the job, and that requirement must be taken into account. It is also important to remember that most printers serve many clients or, in the case of in-house printers, many departments.

Desktop Publishing

Many firms have facilities for producing camera-ready copy with laser printers or other high-quality printing devices that accept computer disks. Some operators of this equipment are able to produce excellent page layouts, including most drawings.

The need for planning is not eliminated by desktop publishing equipment, but the time, effort, and coordination that would have to be otherwise taken into account may well be reduced. If you are fortunate enough to work for a firm that has such facilities, learn all you can about them. In all instances, knowledge of the entire document production process is essential for planning any writing job.

Specifications and Style Guides

Much technical writing is done in accordance with specifications. For example, a contract that requires a final report as part of a development project will probably specify that the report be written and produced in accordance with a certain specification. (Specifications themselves are examples of technical writing, but this section does not deal with specification writing. Specifications typify instructional writing, as described in Chapter 12.) Specifications can, and often do, cover every aspect of a document's preparation, from writing style to reproduction process. They can be either a help or a hindrance, but they cannot be avoided. Unfortunately, a specification whose purpose is to tell a writer how a document must be written doesn't always itself exhibit the principles of good writing. Nevertheless, specifications are a necessary evil, and every technical writer has learned to live with them.

One very important fact must be taken into account at the outset of any technical writing project: *All matters of format prescribed by specifications are arbitrary.*

There are, quite literally, scores of ways to format a document. There is no one right way to do the job — except the way you have been instructed to do it in any given instance. In fact, many of the conventions dealt with in style guides are also often peculiar to a certain field of endeavor or to a certain company. Bibliographic reference style is one such convention that varies greatly with locale and area of interest. The Modern Language Association gives its "right" way of citing references. And just about every professional body in every field has its "right" way, as well.

Knowing that this situation exists can prevent you from making a fool of yourself by "correcting" someone else's format or mechanical style.

Specifications

A typical writing specification is given in the following sample. Working to this specification now will lessen the difficulty of having to work with one for the first time on the job. This specification may be followed in completing many of the exercises in this book.

SAMPLE 7-1 **A Sample Specification**

SPECIFICATION PEL-R-51733

1. SPECIFICATION COMPLIANCE
    1.1. This specification supersedes all previous issues.
    1.2. In case of a conflict in matters of mechanical style between this specification and other relevant documents, this specification shall govern.

2. FORMAT
    2.1. Heading Style
        2.1.1. Sequenced alphanumeric designators shall be assigned to each heading. A Roman numeral shall be used for each primary heading; a capital letter shall be used for each second-order heading; an Arabic numeral shall be used for

continued ➔

---

each third-order heading; a lower-case letter shall be used for each fourth-order heading; an Arabic numeral enclosed in parentheses shall be used for each fifth-order heading; a lower-case letter enclosed in parentheses shall be used for each sixth-order heading; seventh-order and eighth-order headings shall be the same as fifth-order and sixth-order headings respectively, except that they shall be underscored also.

2.1.2. Primary headings shall be in all capital letters and shall be underscored. Secondary headings shall be in initial capital letters and shall be underscored. All other headings shall be in initial capital letters and shall not be underscored.

2.1.3. Primary headings shall not be indented. Secondary and subsequent headings shall be indented in successive five-space intervals. Paragraphs shall not be assigned an alphanumeric designator. The first line of each paragraph appearing beneath a heading shall be aligned with the first letter of the heading. Second and subsequent lines shall be aligned with the left-hand margin. A sample sequence is shown below:

[START OF SAMPLE SEQUENCE]

I. SYSTEM DESCRIPTION

   A. Special Features

      1. Multiple-Function Option

      The multiple-function option is a desirable feature if the unit is to be used in a closed environment ...

      2. Easy Maintenance

      Maintenance must be carried out on a regular basis. The maintenance schedule must include humidity-damage assessment and periodic cleaning.

         a. Humidity Damage

         Inspection for and repair of humidity damage is essential to the life of the equipment ...

         b. Cleaning

         Cleaning shall be accomplished at established intervals ...

            (1) Cleaning Methods

            Several cleaning methods are approved for this ...

continued →

(2) Cleaning Agents

Cleaning agents shall conform to the requirements of Fed-D-121-1.

(a) Soaps

The soaps approved for use include those that ...

(b) Other Agents

The agents approved for use in cleaning of ...

(1) Types

Two types of agents are to be avoided. Those agents are ...

(2) Additives

Ammonia and chlorine may be ...

(a) Ammonia

Solutions with a concentration of ammonia greater than ...

(b) Chlorine

Solutions with a concentration of chlorine greater than ...

II. FUNCTIONS AND CONTROLS

All functions and controls have been engineered for maximum utility. The vernier control devices have been ...

[END OF SAMPLE SEQUENCE]

2.2. Illustrations

2.2.1. Illustrations shall be designed as partial-page presentations whenever possible. This arrangement will often permit the inclusion of illustrations on the same pages on which their reference appears. Placement of illustrations as near as possible to and following their respective references is desirable in all cases. When a full-page illustration is necessary, it shall be placed on the page following the page on which its reference appears.

2.2.2. Foldout pages shall not be used unless no alternative exists. Pages larger than 11×17 inches in size are not approved for use in any case without prior authorization.

2.2.3. Each illustration shall be assigned a figure number, and reference to an illustration shall be made only by its fig-

continued ➤

ure number. Figure numbers shall be assigned in sequence, according to the order of appearance of the illustrations throughout the text. Arabic numbers shall be used.

2.2.4. A caption, composed of the word *figure,* the numeral assigned, and an appropriate title, shall appear beneath each illustration appearing in the text.

2.2.5. All line representations, including drawings, diagrams, charts, plots, curves, schematics, photographs, and lithographs, shall be regarded as illustrations.

2.3. Tables

2.3.1. Tables shall be designed as partial-page or full-page presentations as required. Each column within a table shall have an appropriate heading. At the discretion of the writer, tables shall be blocked-in, and columns shall be separated by lines.

2.3.2. Like figure numbers, table numbers shall be in Arabic numerals, but the sequence shall be separate and distinct from figure numbers.

2.3.3. Like illustrations, tables shall be identified by captions. Each table caption shall be composed of the word *table,* the numeral assigned, and an appropriate title. Table captions shall appear at the top of the table. If a table must be continued on a second page, it shall be identified in the following manner: "Table 1, continued."

2.3.4. No table shall appear before a reference to its table number has appeared in the text. Each table shall be placed as near as possible to its reference.

2.4. Pagination

2.4.1. All pages of the report (beginning with the title page) shall be numbered consecutively in Arabic numerals. The title page shall bear the notation *Page 1 of (total) pages* in the upper right-hand corner.

2.4.2. Each succeeding page number shall be centered 1 inch from the bottom of its respective pages.

2.5. Margins

2.5.1. The left-hand margin shall be a minimum of 1-1/2 inches. The right-hand margin shall be a minimum of 1-1/4 inches.

continued →

2.5.2. The top margin shall be at least 1-1/2 inches. At least two line spaces shall be allowed between the last line of text and the page number.

2.6. Numerals in Text

Numerals shall conform to the standards established in an appropriate style guide [such as the one contained within the discussion following this specification].

2.7. Copy Quality

2.7.1. All reports shall be typewritten or produced by a good quality printer. Dot matrix that approximates letter quality is acceptable.

2.7.2. Single spacing shall be used, in order to provide an accurate simulation of report style.

3. CONTENT

The report shall comprise the front matter, the report proper, and appendices as required.

3.1. Front Matter

Front matter shall include a cover, a letter of transmittal, a title page, an abstract, a table of contents, a list of illustrations, and a list of tables.

3.1.1. The cover shall be of the duo-tang type and shall bear the title of the report, the date, and the author's name.

3.1.2. The letter of transmittal shall be in accordance with good technical writing practices. [See the directions for writing a letter of transmittal in Chapter 10.]

3.1.3. The abstract shall present a brief summary of the contents of the report. The word *ABSTRACT* shall appear at the top of the page. The abstract shall not exceed one page in length. [See the directions for writing an informative abstract in Chapter 10.]

3.1.4. The table of contents shall list the front matter (excluding the cover, the letter of transmittal, and the table of contents), all headings in the report, and all appendices, keying each to a page number.

3.1.5. The list of illustrations shall list the figure number, the title, and the page number of each illustration appearing in the report.

continued ➤

3.1.6. The list of tables shall list the table number, the title, and the page number of each table appearing in the report.

3.2. Report Proper

3.2.1. All headings shall appear in the report in the style mentioned in Section 2.1., Heading Style.

3.2.2. The writing style shall conform to the standards of good technical writing practice.

3.2.3. Spelling shall be in accordance with a standard dictionary, such as the Canadian Edition of The New Lexicon *Webster's Encyclopedic Dictionary of the English Language.*

3.2.4. Abbreviations standards are many and varied. The report shall observe abbreviation standards that conform to good technical writing practice. [The student may observe the standards set forth in the sample style guide appearing later in this chapter.]

3.2.5. Appendices shall be included as necessary. Page numbering shall be continued through all appendices so that the page count total on the title page will not be compromised.

Style Guides

Companies often have their own style guides. The advantages of having an in-house style guide are several. First, some contracts will not indicate specifications to be followed in technical documents; instead, these contracts use the phrase "in accordance with good commercial practice" or words to that effect. Also, the in-house style guide makes for uniform internal documents. Finally, authorities do not agree on writing forms and conventions, and an internal style guide can serve as a final reference for all disputed matters.

Although a company style guide might deal with such topics as cover design, letters of transmittal, abstracts, etc., the following sample style guide omits these items because they are covered separately in Chapter 10. Company style guides also provide information in matters of mechanical style, such as abbreviations, punctuation, and spelling, etc.; these items are detailed in the following sample. This style guide may be followed in completing the exercises contained in this book.

PELLSTON CORPORATION STYLE GUIDE

INTRODUCTION

This style guide has been prepared so that employees of the Pellston Corporation responsible for technical documentation will produce uniform reports, proposals, and manuals.

It should be noted here that no single style guide can govern all documents. Customer requests for you to follow certain specifications, the customer's company style guide, or other conventions must at all times be honored. However, in the absence of specified documents, this style guide may be used.

APPROVED USE OF ABBREVIATIONS, SYMBOLS, AND ACRONYMS

First, distinctions must be made among abbreviations, symbols, and acronyms.

ABBREVIATIONS

An abbreviation is a shortened version of a word. Abbreviations can be very useful, particularly within tables, on drawings, or in any other place where space is short. However, abbreviations (with a few exceptions) are out of place within text. The exceptions are standard publishers' abbreviations, such as e.g. and i.e., and abbreviated units of measure preceded by figures. However, note that units of measure preceded by spelled-out words must be spelled out also.

A writer using abbreviations must be careful to use only long-established forms. A major shortcoming of abbreviation usage practices is that there is no widespread agreement on forms. Specifications governing abbreviations differ greatly. When an abbreviation specification is in effect for a particular document, your job as a writer is simplified; that is, you must follow the specification. However, when a choice does exist, you must consult the authority that has most accurately recorded the abbreviation practices of the particular field for which the document is being prepared.

The attempted imposition of overall standards by governmental or other semi-official bodies has been, paradoxically, a serious impediment to standardization. Forgetting that language can't be legislated successfully, one or another of these bodies often tries to do exactly that. One publication alone lists more than 7000 abbreviations, spanning a wide range of scientific and engineering disciplines. Unfortunately, the consensus claimed by the

continued ➜

originators of this book is but an agreement among a very small and non-representative group of people. This book would be a great step forward if everyone in the scientific community could be forced to use it. However, such is not the case.

SYMBOLS

Although symbols are not the same as abbreviations, non-experts in the use of English often confuse the two. This tendency to confuse abbreviations and symbols has come about because the authorities at Système International d'Unités (SI) decided on Roman letters as the stuff of many of their symbols.

The confusion thus produced has led to non-standard spellings, which Pellston employees should diligently avoid. In essence, these misspelled words are back-formations arrived at from the false belief that letter symbols are shortened versions of the words to which they refer. These errors include words such as kiloHertz (back-formed from the SI symbol kHz). Kilohertz is English. KiloHertz is not.

Pellston personnel are directed to distinguish between abbreviations and symbols to avoid this kind of error from occurring in company documents.

ACRONYMS

An acronym is an abbreviation that is pronounced. Usually, an acronym is formed by the first letter of each word of something, such as LED for light emitting diode. Eventually, nearly everyone forgets what the acronym stands for, and the acronym itself becomes a word. Some examples are Radar, Laser, UNICEF, and NATO.

SUMMARY

Do not use an abbreviation in text unless it meets long-established publishing standards or unless it refers to units of measure and is preceded by a figure.

Use common acronyms if they are better known than their derivations. If the derivation of a lesser known acronym or abbreviation is unwieldy and if the shortened version is to be used repeatedly, spell out the word or words in full the first time, placing the acronym or abbreviation in parentheses following it; use the shortened version thereafter. Select abbreviations in accordance with their acceptance in the technical field in which you are working.

ORGANIZATIONS AND DOCUMENTS TO CONSULT

- American Technical Society
- American Chemical Society
- American Institute of Physics

continued →

- Institute of Electrical and Electronics Engineers
- *The Canadian Style: A Guide to Writing and Editing* (the style guide used by the federal government)
- *Webster's Third New International Dictionary of the English Language, Unabridged*

NUMBER USAGE GUIDE

The general rule governing the choice between numbers expressed as spelled-out words or figures is that spelled-out words are used to denote numbers less than 10, and figures are used to denote all numbers of 10 or greater. However, traditional technical writing practice is to write figures rather than spelled-out words to denote numbers less than 10 in the following instances:

- in statements in which more than one number is mentioned (example: The inventory includes 3 guitars, 1 banjo, and 3 violins.)
- in a document in which numbers are used extensively
- preceding symbols or abbreviated units of measure
- when expressing decimals or fractions
- in tables or illustrations

Of course, a number that begins a sentence must always be spelled out. If the number is very large, recast the sentence so that the number falls elsewhere in the sentence and express it as a figure.

Decimals less than 1 should be written with a 0 preceding the decimal point. Numbers greater than 1 million should be written with figures preceding the appropriate word, for example, 10 million, 4 billion, etc. Scientific notation may also be used if its use is appropriate to the situation. The following examples illustrate correct use:

The container could hold 4×10^9 units of the gas.

The cost of the containers exceeded $4 billion.

Note also that the dollar symbol is used to precede figures that refer to dollars.

Do not use two figures in succession. Either rewrite the sentence or spell out one number, as in the following example:

The equipment included four 100-liter containers.

PUNCTUATION

Punctuation should follow the standard rules of formal writing. Consult an authoritative guide. [See, for example, the Self-Directed Course in English Grammar and Usage, Section V of this book.] Beware of those authorities

continued →

who cite exceptions to the rules as though the exceptions themselves were also rules. Avoid those reference guides that offer alternatives without giving the stylistic differences inherent in the choices, merely stating that certain rules may sometimes be disregarded. These authorities should not be consulted, for they introduce an uncertainty that places an unnecessary burden on the technical writer. Worse, such "non-rules" create an environment conducive to ambiguous statement, which has no place in Pellston's documents.

SPELLING

Spellings should be in accordance with an up-to-date Canadian dictionary such as the Canadian Edition of The New Lexicon *Webster's Encyclopedic Dictionary of the English Language.*

Exercises

7-1. Following a Specification

1. Study Sample 7-1.

2. Select any piece of writing, preferably something you already have on your computer hard drive or on a floppy disk, and apply the heading and paragraphing scheme given in Sample 7-1. If your word processing program permits, use the style feature in setting up your document.

3. If the piece you have selected has illustrations, also follow the figure reference and caption scheme specified in Sample 7-1.

7-2. Following a Style Guide

1. Study Sample 7-2.

2. Select a piece of technical writing and edit it to make sure all abbreviations, acronyms, and symbols conform to the style specified in Sample 7-2.

3. Edit the same piece for number usage, using the style guide in Sample 7-2.

4. By referring to the punctuation rules given in Section V of this book, A Self-Directed Course in English Grammar and Usage, edit your selection for punctuation.

CHAPTER 8

Illustrations and Tables

OBJECTIVES: In this chapter, you will learn that the spatial and parallel modes of exposition are best presented in illustrations and tables. You will also learn how to choose the form of graphic illustration and assemble tables to make the reader's job easier than it would otherwise be.

Graphic Illustration

Illustration Types

Illustrations include photographs, graphs, charts, diagrams, and line drawings of all kinds. As noted elsewhere in this book, standard technical writing practice identifies all the different spatial representations of information as one broad category called illustrations. Therefore, when you design a technical document, you will have one sequence of figure or illustration numbers, not one for photographs, another for diagrams, charts, etc.

Sometimes, the choice of an illustration type is obvious. For example, you would choose a block diagram to show a company organization (see Figure 8-1).

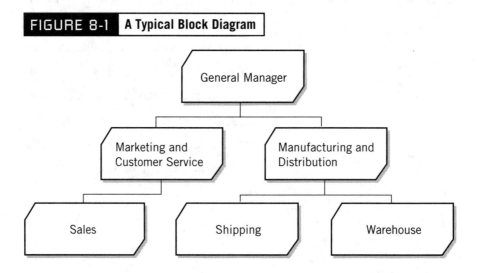

FIGURE 8-1 | **A Typical Block Diagram**

At other times, several options may be available, and technical writers must choose illustrations that best suit their purposes. Figure 8-2 shows an illustration chosen by the author of a paper discussing the effects of a particular pilot training technique. Charts such as this one may be computer-generated using a program such as Microsoft Excel.

The author could have chosen to show the data in a table since a comparison of accident rates in two countries is being presented. However, a spa-

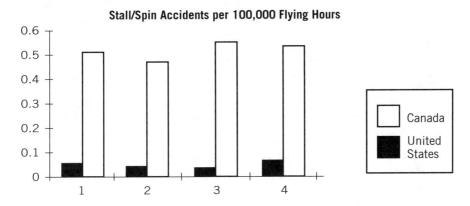

FIGURE 8-2 A Typical Column Chart

tial presentation of the data shows the differences much more effectively than would a simple parallel (side-by-side) placement of numbers.

The author could also have chosen a line chart, Figure 8-3, or a bar chart, Figure 8-4 (both of which could be generated by a program such as Microsoft Excel). However, the line chart shows only that the rates both remain low, comparatively speaking, over time when the important point is that the rate of occurrence of this type of accident is ten times higher in Canada than it is in the United States. Had the variation over time been important, this chart might have been a good choice.

The bar chart would also be a pretty good choice although the column chart is probably the better of the two.

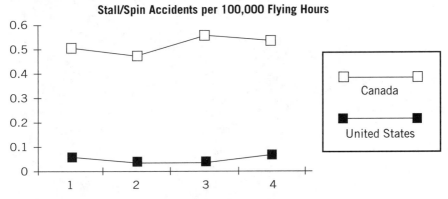

FIGURE 8-3 A Typical Line Chart

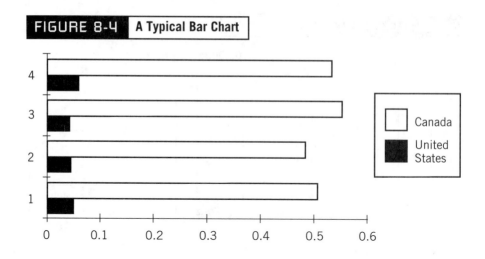

FIGURE 8-4 | A Typical Bar Chart

Canada

United States

Photographs are often useful in showing certain kinds of information. Figures 8-5, 8-6, and 8-7 show several steps in a disassembly sequence. This sequence would, of course, also rely on the technician's following a series of written-out steps. However, the photographs show the technician exactly how to carry out the operation, even indicating proper hand placement. A line diagram giving the same information would have to be quite complicated.

FIGURE 8-5 | A Disassembly Sequence: Positioning the Retaining Ring Removal Tool

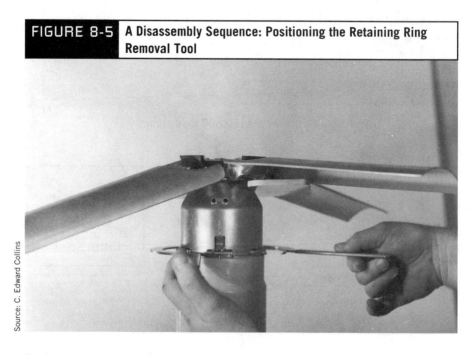

Source: C. Edward Collins

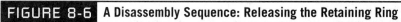

FIGURE 8-6 A Disassembly Sequence: Releasing the Retaining Ring

Source: C. Edward Collins

FIGURE 8-7 A Disassembly Sequence:
Removing the Rotochute

Source: C. Edward Collins

On the other hand, the equipment connections shown in Figure 8-8 would be better shown in a line drawing that includes information to clearly identify the connector points, valve positions, control settings, etc.

FIGURE 8-8 | Making Equipment Connections

Source: C. Edward Collins

Pictorial drawings of various kinds serve varying purposes. Figure 8-9, for example, shows an "exploded view" of an aileron bell-crank assembly (left) along with a view of the same assembly in its assembled state (right). This illustration, which accompanies a series of instructional steps, helps the person assembling the aileron bell-crank to see exactly how all the pieces fit together.

Figure 8-10, an exploded-view drawing of a rudder pedal assembly, has been designed for the same purpose, to help the user put the device together. Since the fully assembled device is more easily discernible in the exploded view, only that one view is required to accompany the instructional steps.

Exploded-view drawings are also sometimes used to identify the parts of an object or to show the relationship of parts to one another in parts catalogues and other similar documents. Used therein, they always include numbers or symbols that are keyed to a parts list. The list may appear as part of the drawing or as a separate table. See Figure 8-11.

FIGURE 8-9 | Two-View Drawing Accompanying an Assembly Instruction

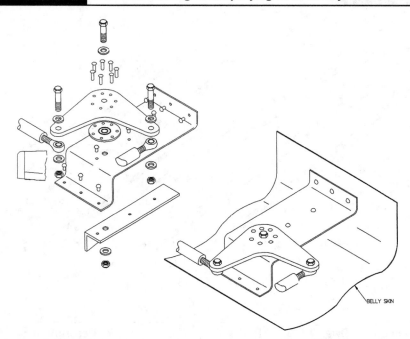

BELLY SKIN

FIGURE 8-10 | Single-View Drawing Accompanying an Assembly Instruction

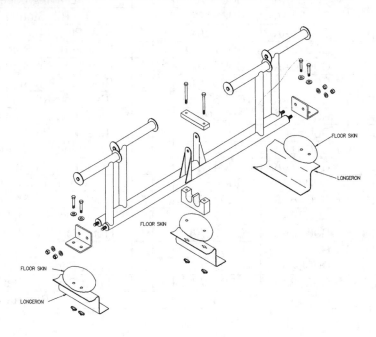

FLOOR SKIN

LONGERON

FLOOR SKIN

FLOOR SKIN

LONGERON

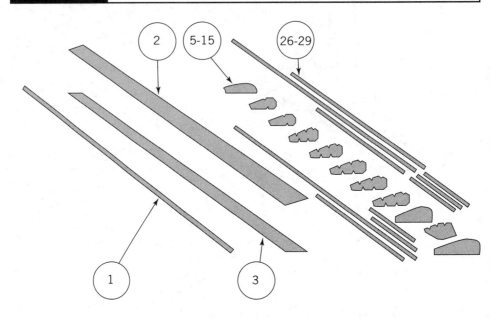

No.	Dwg. #	Part #	Qty.	Description
1	G0094AA	G0094AA	1	Port Rear Spar Assembly 90
2	G0036AA	G0036AA	1	Port Upper Skin 90
3	G0038AA	G0038AA	1	Port Lower Skin 90
4	G0034AA	G0034AA	1	Port Leading Edge Skin
5	G0081AA	G0081AA	1	Port Tip Rib Station 177.75
6	G0079AA	G0079AA	1	Rear Port Rib Station 162
7	G0075AA	G0075AA	1	Rear Port Rib Station 144
8	G0071AA	G0071AA	1	Rear Port Rib Station 126
9	G0067AA	G0067AA	1	Rear Port Rib Station 108
10	G0063AA	G0063AA	1	Rear Port Rib Station 90
11	G0059AA	G0059AA	1	Rear Port Rib Station 72
12	G0055AA	G0055AA	1	Rear Port Rib Station 54
13	G0051AA	G0051AA	1	Rear Port Rib Station 36
14	G0047AA	G0047AA	1	Rear Port Rib Station 18
15	G0043AA	G0043AA	1	Rear Port Root Rib
16	G0078AA	G0078AA	1	Front Port Rib Station 162
17	G0074AA	G0074AA	1	Front Port Rib Station 144
18	G0070AA	G0070AA	1	Front Port Rib Station 126
19	G0066AA	G0066AA	1	Front Port Rib Station 108
20	G0062AA	G0062AA	1	Front Port Rib Station 90
21	G0058AA	G0058AA	1	Front Port Rib Station 72
22	G0054AA	G0054AA	1	Front Port Rib Station 54

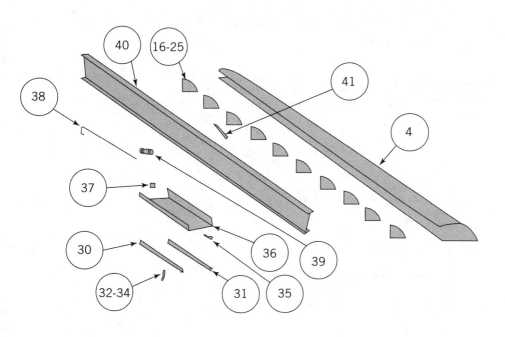

No.	Dwg. #	Part #	Qty.	Description
23	G0050AA	G0050AA	1	Front Port Rib Station 36
24	G0046AA	G0046AA	1	Front Port Rib Station 18
25	G0042AA	G0042AA	1	Front Port Root Rib
26	G0020AA	G0020AA	1	12 ft. Stringer
27	G0021AA	G0021AA	1	9 ft. Stringer
28	G0022AA	G0022AA	1	6 ft. Stringer
29	G0023AA	G0023AA	1	3 ft. Stringer
30	G0025AA	G0025AA	1	Port Aft Tank Bulkhead
31	G0024AA	G0024AA	1	Port Forward Tank Bulkhead
32	G0014AA	G0014AA	1	Fuel Gauge Cover
33	G0016AA	G0016AA	1	Fuel Gauge Glass
34	G0017AA	G0017AA	1	Fuel Gauge Spacer
35	G0030AA	G0030AA	1	Fuel Fittings Bracket
36	G0032AA	G0032AA	1	Tank Basin Port
37	G0031AA	G0031AA	1	Vent Plate
38	G0015AA	G0015AA	1	Vent Tube
39	G0087AA	G0087AA	1	Gas Filler Neck
40	G0082AA	G0082AA	1	Assembled Main Spar Port
41	G0009AA	G0009AA	1	Strut Pick-up Lug

Another common illustration is the "phantom-view" drawing, which permits the viewer to see an assembly or system inside another assembly. Figure 8-12 shows the location of the fuel line within the fuselage of the all-Canadian-built aircraft, the Griffin.

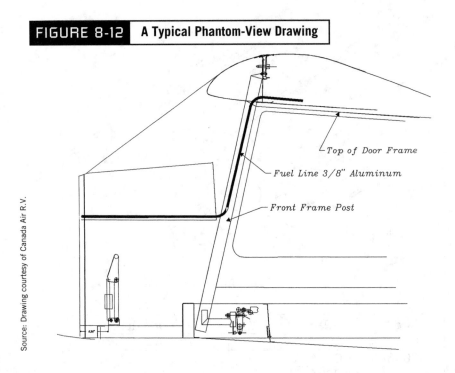

FIGURE 8-12 | **A Typical Phantom-View Drawing**

Source: Drawing courtesy of Canada Air R.V.

─ Top of Door Frame

─ Fuel Line 3/8" Aluminum

─ Front Frame Post

Tables

Tables are useful for the placement of comparative information in a parallel (side-by-side) format. See Table 8-1 at the top of page 131.

Often, tables contain comparisons of the features of competing products, in which case the feature is present in one product and not in another. This kind of table gives the features to be compared as heads of columns, under which either a "yes" or "no" entry of some kind, such as a check mark, is made. Sometimes, instead of a check mark, a comment is entered, as shown in Table 8-2 at the bottom of page 131.

TABLE 8-1	Live Births by Province in 1995		
Province	**Male**	**Female**	**Total**
Newfoundland	2,985	2,874	5,859
Prince Edward Island	898	856	1,754
Nova Scotia	5,422	5,304	10,726
New Brunswick	4,417	4,146	8,563
Quebec	44,782	42,635	87,417
Ontario	75,110	71,153	146,263
Manitoba	8,201	7,912	16,113
Saskatchewan	6,798	6,701	13,499
Alberta	19,878	19,038	38,916
British Columbia	24,194	22,626	46,820

Source: Statistics Canada, Cat. No. 91-002; 91-213

TABLE 8-2	Comparison of Standard and Optional Equipment		
Feature	**Moto-Guzz V-8**	**Flatiron 4**	**Duke GT**
airconditioning	standard	optional	standard
power steering	standard	optional	optional
power brakes	standard	standard	standard
upholstery	vinyl	cloth	leather
power door locks	no	no	yes
outside mirror adjust	power	manual	power
transmission	automatic	4-speed; optional automatic	5-speed

At the author's discretion, the products may appear as column heads, and the features may line the left-hand side of the table. This arrangement can often be the better choice, since most readers scan column heads and glance downwards before looking for specific information. Thus, the reader's natural eye motion will quickly reveal the product with the most standard features. The downside is that the number of columns is limited by available page width, usually about five for a standard page.

Exercises

8-1. Designing an Illustration

Visit three or more grocery stores and record the prices of grapefruit, carrots, celery, 2% milk, top sirloin steak, and canned tomatoes. Present your findings in a column or bar chart, either hand drawn or computer-generated.

8-2. Designing a Table

Assemble the following information into a table.

Model 415 Autopilot Couplers have failed in flight in 10 of the 37 aircraft in which they were installed. The units were all installed in December of last year and, in all cases, failed within the first 2 hours of flight time. Eight of the units failed in the first hour of flight time. The other two failures occurred during the second hour of flight. All the failures occurred in Convair 240 aircraft belonging to Sampson Airlines.

Model 415 Autopilot Couplers have operated successfully for a year in the 27 remaining airplanes, which belong to two other airline companies. Southern Alberta Airways has 21 Model 415 units. Fifteen are installed in their DeHavilland Dash 7s; four units are installed in their Shorts 330s; the remaining two units are installed in their Beechcraft 99s. Both of the Beechcraft 99 installations were done in June, but only 5 of the Dash-7 installations were in June. The Shorts 330s weren't fitted with Model 415s until August (1 unit) and September (the final 3 units). The remaining 10 units were installed in the Dash 7s as follows: July (3 units), August (2 units), and September (5 units).

Northern Manitoba Airlines has 6 Model 415s in operation, installed in their Beechcraft 99 fleet as follows: June (2 units), July (1 unit), and August (3 units). Neither Southern Alberta Airways nor Northern Manitoba Airlines has had any failures.

SECTION III

The
Forms

Introduction

This section includes student exercises based on cases related to a fictional company called the Pellston Corporation. The exercises do not require any technical knowledge. They have been designed so that anyone can successfully complete them.

The Pellston Corporation, although it is purely fictional, is realistically conceived, in order to provide you with a simulated environment that is as close to reality as possible. The situations are typical, and the tasks you are asked to perform are similar to the kind you can expect on the job.

The Pellston Corporation

The Pellston Corporation is located at 4567 Airport Road, Sunnyside, Ontario, K2V 1M8. The president's name is Grant Dobbs. Pellston Corporation's organization chart is shown in Figure III-1.

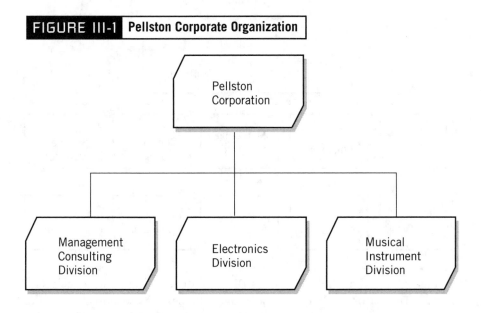

FIGURE III-1 Pellston Corporate Organization

Management Consulting Division

The Management Consulting Division is located in the same building as corporate headquarters. Under the direction of Vice President Manfred Panfreid, it includes 12 specialists in business management, 6 stenographers, 1 technical writer, and 1 desktop publishing specialist. The consultants regularly work in customers' firms, and they are responsible for generating various communications for the customer. They report to the Pellston Corporation.

Electronics Division

The Electronics Division, also located at the same address as corporate headquarters, comprises a full range of design and manufacturing facilities and personnel. Total employment is about 1,500. The organization chart of the Electronics Division is shown in Figure III-2.

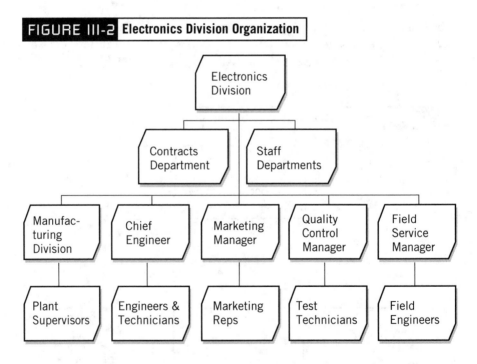

FIGURE III-2 Electronics Division Organization

Please note that the Electronics Division does not include a Technical Writing Department. Chief Engineer Darla Petoskey is responsible for the quality of all contractual reports and handbooks. Therefore, she has had a style guide prepared by Ont-Aero, Inc., to establish writing standards; and she has an up-to-date file of all relevant writing specifications. Large handbook jobs are subcontracted to Ont-Aero, Inc. (Please refer to the sample specification and style guide given in Chapter 7.)

The Marketing Manager is responsible for the quality of all proposals produced in the Division. The style guide used by the Engineering Department is also used in proposal preparation.

Musical Instrument Division

The Musical Instrument Division dates back to 1929 and was originally incorporated as the Happy Music Co. Located at 567 Cherry Blossom Avenue, Surrey, British Columbia, V4D 3Y7, it was purchased by the Pellston Corporation in 1980. The division manufactures stringed instruments only. About 250 employees produce a modestly priced array of guitars, banjos, violins, violas, cellos, and bass viols. Although the instruments produced are not concert quality, they are highly regarded by music teachers and students. The Division is very simply organized. It is managed by Harry Peabody, who has been with the company since 1966, 14 years before Pellston acquired ownership. The Musical Instrument Division is composed of staff and line departments, as shown in Figure III-3.

FIGURE III-3 Musical Instrument Division Organization

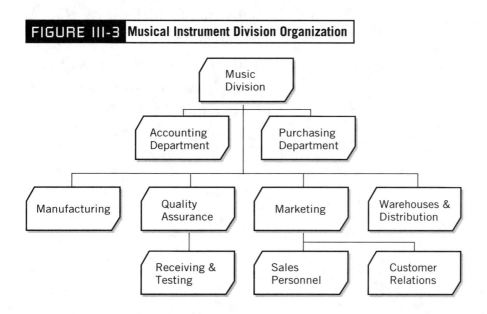

CHAPTER 9

Short Reports

OBJECTIVES: In this chapter, you will learn about the purposes of an internal document system and of the necessity for well-planned and carefully written memorandum reports, proposals, and instructions, all of which may include a heading structure, illustrations, and front matter. You will also learn how the letter format can be used as a form for external documents.

A Word about Report Categories

This chapter is about "short" reports, and Chapter 10 is about "long" reports. This division of the reporting function is convenient for discussion, but it also poses a few questions.

- How short is short and how long is long?
- Is the format of a short report different from that of a long report?
- Are short reports informal and long reports formal?

The answer to the first question is that there is no single answer. Some people regard any report longer than 10 pages as a long report. Others think of any report with headings, illustrations, or front matter as a long report. And then there are those who write 100-page memorandums including all the report elements, and consider them to be short reports.

For the purpose of study, the division is best related to the format, giving us a *yes* to the second question. If the report can be written as a memorandum or a business letter, therefore, this book considers it to be short report, the number of pages and the inclusion of elements such as illustrations, a table of contents, etc., notwithstanding.

The answer to the third question is *no*. What does the word *formal* mean? It means "having form," and all written reports have a form (refer to the discussion of "Formal, Informal, Professional, and Popular Writing" in Chapter 5). The differences in writing style, which are far less distinctive than many people believe, are noted later in this chapter.

Finally, a few popular notions about other report categories need to be discussed. What is a "trip report"? There is no such thing. Employees who travel on business do so for well-defined purposes, and the reports they write are about having achieved or failed to achieve those purposes. The travel aspect is incidental.

Laboratory, inspection, site, and incident reports likewise inhabit the fringes of technical writing activity. Laboratory workers write laboratory reports, most of which consist of filled-in forms; building inspectors write inspection and site reports, and many of those are also filled-in forms. Police officers regularly write incident reports, and (you guessed it) most of those are filled-in forms. Our look at short reports omits a discussion of ways to fill in a form. Forms differ so greatly that no comprehensive information about filling them in could be given. Our discussion is divided into "Internal Documents" written in memorandum format and "Short External Reports" written in letter format.

Internal Documents

The Purpose of an Internal Document System

You will write internal documents for a variety of reasons. For example, you may write weekly, monthly, or quarterly reports describing your progress on the projects on which you are working; you may write reports that will permit your work to be coordinated with the work of other people in the company; you may write internal proposals suggesting new products, new avenues of research and development, or new directions for the work you are doing; or you may write instructions for executing new processes or carrying out company policies. No matter what purpose your documents serve, you will be expected to submit them on time and in a form that is in accordance with company standards.

What Is a Memorandum?

Some people think that a memorandum is a note to oneself. Although its Latin root does suggest that meaning (something to be remembered), memorandums are much more than casual notes on slips of paper. Many different situations call for memorandums. Most internal reports, proposals, and instructions are written in memorandum format, as are routine communications between people working on the same project.

Types of Memorandums

Suppose you are in charge of building the preproduction model of a piece of hardware. You might need to write a memorandum to the purchasing department to indicate what materials and subassemblies you need, a memorandum to the model shop to request services or equipment, and perhaps a memorandum to your supervisor to set out your plan for doing the job. Having completed the work, you might write another memorandum to your supervisor as a final report.

Memorandums that do not fall into one of the general categories of technical writing, namely, proposals, reports, and instructions, roughly parallel external letters of request and responses to requests described in Chapter 13.

Memorandum Format

The memorandum, which may also be called *internal correspondence, interoffice memo, interdepartmental letter,* or something similar, is usually written on a preprinted sheet. The purpose of preprinting is to relieve the writer of having to bother with a letter format. One merely fills in the blanks and then proceeds with the message. Sample 9-1 shows a memorandum heading format typical of those found in many business firms.

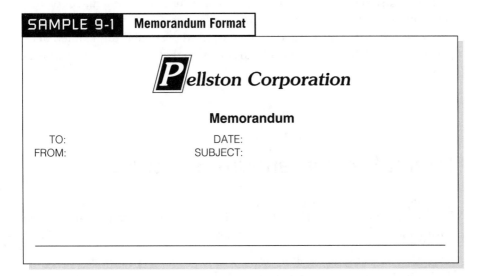

SAMPLE 9-1 **Memorandum Format**

Pellston Corporation

Memorandum

TO: DATE:
FROM: SUBJECT:

Memorandum Style and Tone

The memorandum form permits a free use of the first person, since it is, in fact, a letter from one person to another. However, one must avoid falling into a narrative mode of expression. In other words, remember that you are relating information, not telling a story.

A memorandum can be any length. Some memorandum reports and proposals are quite complex, containing illustrations and tables, and exhibiting a heading structure either with or without an alpha numeric system of designation. Such memorandums should obey all the rules of good publications practice. That is, illustrations should be numbered consecutively and referred to in the text before they are inserted. Tables should be handled in like manner.

As a matter of courtesy and custom, most people either sign or initial a memorandum after completing the message. Some companies prefer that memorandums be signed. Others consider initials to be appropriate, since the name and position of the sender is normally indicated after the "FROM:" line at the top. The practice in some companies has the writer initial the memorandum in the heading, next to the name printed on the "FROM:" line, and the practice in other companies has the writer initial the memorandum at the end. Neither a salutation nor a complimentary close is included as part of a standard memorandum format. However, if the company for which you work has established practices that contradict standard memorandum format, you must follow company policy. Memorandum format is a matter of local convention, not of universal rules and regulations.

Occasionally, the memorandum is considered unsuitable for an internal report or proposal. In these cases, a form similar to that of an external document may be used.

Internal Memorandum Structure

A memorandum longer than five pages should contain headings, so that a framework is visible for quick reference. It is also a good idea to begin with an abstract if the memorandum is more than 10 pages long.

Note: *Abstract* is the term often used to describe a brief summary placed at the beginning of a technical document. Abstracts, summaries, executive summaries, and other front matter items are discussed at length in Chapter 10.

A memorandum longer than 10 pages should have, in addition to an abstract, a table of contents listing the headings and their respective page numbers.

When writing any memorandum, you must bear in mind that its immediate use is not its only use; a memorandum is a permanent record of the internal transactions of a firm. Such consideration will affect what you write and the way you write it. Consider Sample 9-2.

MEMORANDUM

TO: Garth

FROM: Barney

DATE: January 25, 1998

SUBJECT: A/P Coupler Project

CBI has done it again. I sent back the latest shipment. Should I go ahead with the idea we talked about the other day? Sparky might be able to do better, but this is a pretty big order. Let's do lunch on Friday.

B.C.

Garth (whoever he is — the writer has not told us his last name or title) may well know what Barney (whoever he is) is talking about, but someone who wants to find out why the project was overbudget will have to track down the mysterious Barney and ask him what this memorandum means. Sample 9-3 is a properly written version of the same memorandum.

MEMORANDUM

TO: Garth Teasdale
General Manager

FROM: Barney Cavendish
Quality Control
Manager

DATE: January 25, 1998

SUBJECT: A/P Coupler Project

As you know, all our circuit boards for the A/P Coupler Project are made by Circuit Boards, Inc., under contract Pel-89-63. Up to now, we have not believed it necessary to test the incoming boards before they go into assembly.

However, many of the boards in the last two shipments have been improperly etched, thereby causing a 17 percent rejection rate in the final-

continued ➔

inspection-and-test stage. This figure is well above the normal rejection rate of between 2 percent and 3 percent.

Should I go ahead with our plan for setting up an inspection program for incoming boards? Catching these defective boards before they go into assembly will save us a lot of expense. However, if more than 12 percent of them have to be rejected, we won't have enough to keep our assembly line in full-time operation anyway.

The only alternative I can think of would be to transfer the contract to Sparky Electronics. Although we would have no trouble in revoking CBI's contract (they agreed to a 99 percent reliability figure), we might experience difficulty in getting the needed production out of Sparky.

As you will recall, we noted that the production space and equipment available was marginal when we visited Sparky Electronics before deciding on a subcontractor for the boards.

I have returned CBI's last shipment for inspection and re-work. We will have to make a decision on this matter before February 14, at which time production will have to stop altogether for lack of circuit boards.

B.C.

The revised example provides all the necessary information for anyone who needs to understand what has happened and what options are being considered.

An example of a complex memorandum report is shown in Sample 9-4.

SAMPLE 9-4 | **A Memorandum Report with an Abstract, a Heading Structure, and Illustrations**

MEMORANDUM

TO: Garth Teasdale
General Manager

DATE: June 10, 1997

FROM: Donna Nebbish
Project Engineer

SUBJECT: A/P Coupler Subcontractor
Selection

continued →

ABSTRACT

My survey of the potential subcontractors for printed circuit board production is now complete. Either Circuit Boards, Inc., or Sparky Electronics appears to be a possible choice. This conclusion is based on their proposals and all other information gathered to date. I recommend that we convene a team to visit CBI and Sparky to have a first-hand look at their facilities. This report includes evaluative comments on each of the proposals received in response to our request for proposal.

OKAY PRINTED CIRCUITS, INC.

Okay has indicated that they have complete facilities at our disposal, and they have assured us that timely delivery can be made of the printed circuits called for in the request for proposal.

Okay's PC lab layout is shown in Figure 1.

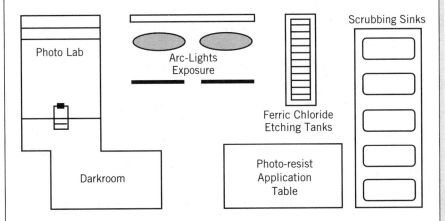

Figure 1 — Okay's Printed Circuit Board Production Laboratory

Okay didn't put any dimensions on their PC Lab, but I phoned Laura Gates in Okay's Engineering Department, and she told me they have about 900 square feet. I don't believe Okay has a full appreciation of the magnitude of this job. Their thinking seems to be that they can hire a few more people and double their production without expanding their operation in either space allotment or capital outlay. A further indication of Okay's naiveté is that their price is less than half that of the other bidders; in other words, Okay would barely recover their material costs.

MIRACLE HOLDINGS, INC.

Miracle Holdings didn't submit a proposal. Instead, they wrote a lengthy letter describing their desire to get into the PC board business. They seem will-

continued ➤

ing to do whatever is necessary, but they don't have any experience in this area. They expect us to consider their participation on a cost-plus-fixed-fee basis, the reason being that they don't have any idea of the costs involved in such a production run. They would have to set up an operation to make the boards. They have space, but that seems to be their only real asset.

SPARKY ELECTRONICS, INC.

Sparky Electronics has been a reliable supplier of subassemblies for us in the past. They have made their own circuit boards for the past five years, and all the subassemblies we have purchased from them have included their own boards. Their record as a supplier of good quality products is untarnished. Figure 2 shows the PC lab layout at Sparky. The production capacity appears marginal for the size of the production run we need, particularly when one considers that the lab's time to be devoted to our project is only 75 percent. Their bid was $1.12 per board.

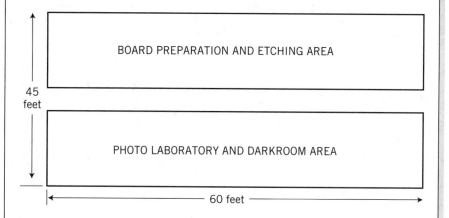

Figure 2 — Sparky Electronics PC Board Laboratory Layout

CIRCUIT BOARDS, INC.

This company is new to the area. It is a branch of Peerless Products Corporation located in Scarborough, Ontario. Their facility is by far the best equipped and most spacious of the four companies responding to our request. The company's only product is circuit boards, and they have set up the company to compete only in that area. For these reasons, this company appears to be a good choice.

They have only one other contract so far, and that one is with Aerostar Unlimited. CBI has been in business for only three months, but they have the backing of the parent corporation and appear very solid indeed. In addition,

continued →

CBI's bid was the best, at $1.10 per board. A floor plan of their plant is shown in Figure 3.

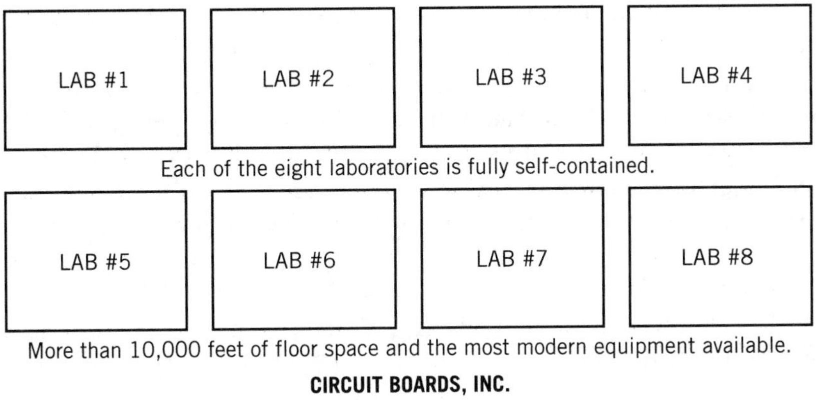

LAB #1	LAB #2	LAB #3	LAB #4

Each of the eight laboratories is fully self-contained.

LAB #5	LAB #6	LAB #7	LAB #8

More than 10,000 feet of floor space and the most modern equipment available.

CIRCUIT BOARDS, INC.

Figure 3 — Floor Plan of Circuit Boards, Inc.

SUMMARY

Of the contractors who submitted serious proposals, CBI's bid is the lowest, and they will have no problem in fitting our project into their production schedule. The downside is that they are untried. Sparky Electronics is a long-time supplier, and their bid is insignificantly higher. The potential problem with Sparky is their busy production schedule. Neither Okay Printed Circuits nor Miracle Holdings should be considered as a serious contender for this contract. Could you, Jill Barnaby (the contracts administrator), and I get together on a decision before June 30?

Thanks, Garth.

D. N.

Short External Reports

The Letter Report

When a short report is submitted as part of a contractual obligation, the contract usually specifies the form the report should take. When a short report is not required by contract or when the contract does not cite a specification that will govern report preparation, it may take the form of an ordinary business letter.

When writing a letter report, you must resolve a minor conflict that exists between the character of a report and the fundamental nature of a letter. The ideal report is objective and impersonal, while a letter has a person-to-person nature. The resolution of this problem is to be found in strict adherence to a standard letter format. The style and tone of the letter report can be impersonal and objective, but if the writer carefully attends to all the format items, the document will retain a personal character.

Letter Format

Two letter formats are common in North America. They are the full-block and the semi-block. Either is generally acceptable, but most firms stick to one or the other. Although you may encounter variations on these two formats, you should understand that such variations are not *generally* acceptable. The authorities who have tried to advance their own preferences in letter format have met with little success.

The conventions of letter format have changed somewhat over the last 100 years, but most of the currently accepted changes have been in response to new technology that is now old, namely, the typewriter. In the meantime, the best advice is probably to stick to the two formats given below for correspondence that will travel through the mail or by courier. Most people using e-mail have already broken away from these conventions, and you can safely do the same.

The parts of the letter are as follows:

- The inside return address
- The dateline
- The inside address
- The salutation
- The body of the letter
- The complimentary close
- The signature group
- Miscellaneous options and notations

Before beginning a study of the parts of the letter, you should know a few things about the appropriateness of abbreviations within letter format. In general, abbreviations are inappropriate within standard format items. If the company for which you work has established a format that includes abbreviations, then by all means use them. In the absence of such established practices, however, you should avoid abbreviations in the inside return address, dateline, and inside address. One obvious exception to this advice is to retain abbreviations that are a part of a company's legal name, such as *Inc.*, *Ltd.*, and *Corp.* The other exception is for provinces and states. The U.S. Post Office has published a list of two-letter abbreviations for the 50 states and possessions and for the 12 Canadian provinces and territories, and their use has become widely accepted in both Canada and the United States. See Table 9-1. Again, however, if you or your company has an established practice different from this method of handling addresses, by all means use it.

THE INSIDE RETURN ADDRESS

The inside return address is the address of the sender of the letter. Nearly all business firms have this information preprinted on their stationery. When preprinted stationery is used, therefore, the inside address is omitted, and the first line of the letter is the dateline.

If plain paper (rather than preprinted letterhead stationery) is used, the inside return address makes up the first group of lines in the letter. The inside return address begins with the street address, *not* with the name of the writer. The second line cites the city, province, and postal code. Some writers separate the name of the city from the province or state with a comma; others do not. The practice in this book is to omit the comma.

In the following example, note that no punctuation is used at the ends of the lines and that the word *Street* (which is part of the name of the street) has its first letter capitalized.

> 4321 Barrymore Street
> Valleyview BC V8N 2U2

THE DATELINE

Two ways of writing the date are acceptable. The traditional way includes the name of the month expressed in its fully written-out form, the day expressed as a figure, and the year expressed as a figure. A comma is used to separate the figures expressing the day and the year. The following example shows this method:

> February 5, 1999

The second option is to place the elements in the sequence of day, month, and year. In writing the date this way, the comma is not needed, as the following example shows:

> 5 February 1999

TABLE 9-1	Abbreviations of Canadian Provinces and Territories, and American States and Possessions		
Alabama	AL	Nevada	NV
Alaska	AK	New Brunswick	NB
Alberta	AB	New Hampshire	NH
American Samoa	AS	New Jersey	NJ
Arizona	AZ	New Mexico	NM
Arkansas	AR	New York	NY
British Columbia	BC	Newfoundland and Labrador	NF
California	CA	North Carolina	NC
Canal Zone	CZ	Northwest Territories	NT
Colorado	CO	Nova Scotia	NS
Connecticut	CT	Ohio	OH
Delaware	DE	Oklahoma	OK
District of Columbia	DC	Ontario	ON
Florida	FL	Oregon	OR
Georgia	GA	Pennsylvania	PA
Guam	GU	Prince Edward Island	PE
Hawaii	HI	Puerto Rico	PR
Idaho	ID	Quebec	PQ
Illinois	IL	Rhode Island	RI
Indiana	IN	Saskatchewan	SK
Iowa	IA	South Carolina	SC
Kansas	KS	South Dakota	SD
Kentucky	KY	Tennessee	TN
Louisiana	LA	Texas	TX
Maine	ME	Utah	UT
Manitoba	MB	Vermont	VT
Maryland	MD	Virgin Islands	VI
Massachusetts	MA	Virginia	VA
Michigan	MI	Washington	WA
Minnesota	MN	West Virginia	WV
Mississippi	MS	Wisconsin	WI
Missouri	MO	Wyoming	WY
Montana	MT	Yukon Territory	YT
Nebraska	NE		

Note that no punctuation is used at the end of the line, no matter which option is chosen.

Unless you want to risk being misunderstood, you should not use figures for all three elements of the date. The date 2-5-1999 can be taken to mean February 5, 1999 or May 2, 1999. The argument that everyone knows the day comes first or that if you write 1999-2-5, everyone knows 2 stands for February is not convincing.

THE INSIDE ADDRESS

Although an inside address is usually not included in a personal letter, its inclusion in a business letter is required. Business letters are filed; the envelopes in which they arrive are discarded. Therefore, the inside address provides the only permanent record of the addressee of a given letter.

A writer of a letter faces several options in addressing a letter. Some of these options are shown in the following examples:

Mr. Julius T. Finch
General Manager
Fantastic Forklift Corporation
9876 Ridgeway Avenue
Calgary AB T7M 4B4

General Manager
Fantastic Forklift Corporation
9876 Ridgeway Avenue
Calgary AB T7M 4B4

Fantastic Forklift Corporation
9876 Ridgeway Avenue
Calgary AB T7M 4B4

Attention: Mr. Julius T. Finch
          General Manager

Fantastic Forklift Corporation
9876 Ridgeway Avenue
Calgary AB T7M 4B4

Attention: General Manager

In the above examples in which a person's name is included, please note that the title of the person appears on a separate line. A variation on that form follows:

Mr. Julius T. Finch, General Manager
Fantastic Forklift Corporation
9876 Ridgeway Avenue
Calgary AB T7M 4B4

Note that a comma separates the person's name and title if both are included on the same line.

The first line of the inside address denotes the addressee. When the addressee is a person (rather than a company), correctly handled mail will not be opened by anyone other than that person, although some companies ignore the distinction. When the addressee is the company and when the person's name appears in an attention line, the letter can be opened by someone other than the person mentioned in the attention line. Knowledge of these conventions will permit you to choose the most desirable method to use in addressing any letter.

When addressing a woman, you will face another choice. Until the late 1700s, women of every age and marital status throughout the English-speaking world were referred to as *Mrs.* Around the end of the eighteenth century, distinguishing between unmarried and married women by referring to them as *Miss* or *Mrs.* became a standard practice.

Everyone is familiar with the attempt to change the standard once again through the institution of the *Ms.* title. Government offices and many businesses now follow the practice of addressing all women as *Ms.* In spite of the efforts of these organizations, however, many people continue to use the *Miss* and *Mrs.* titles.

There are probably at least two reasons that the practice remains mixed. First, surveys have shown that efforts to impose the *Ms.* title have led to opposition and confusion. Some women, both married and single, object to the title, and surveys have shown that many people believe it appropriate only for divorced women. The second reason is that, as history tells us, attempts to impose language practices seldom achieve complete success — at least not right away. Perhaps in a generation or two, *Ms.* will have become the standard its advocates hope to create.

In the meantime, the sensitive letter writer attempts to discover the title preferred by the addressee, rather than to risk offending someone.

THE SALUTATION

The salutation is a way of getting into a letter gracefully. When you write *Dear Mr. Smith* at the beginning of a letter, Mr. Smith will not for one moment think that he is dear to you. In fact, such a salutation is effectively invisible for the simple reason that the reader expects it.

On the other hand, when you write *Mr. Smith*, Mr. Smith might take offense for the same reason: that is, the salutation does not meet his expectations and thus becomes visible, leading him to subconsciously wonder why you didn't address him as *Dear Mr. Smith*.

Fifty years ago, including the addressee's name in the salutation was considered unacceptable unless the writer had been personally introduced to the addressee. However, this restriction is no longer observed by very many people. Those who do observe it use *Dear Sir, Dear Madam,* or *Gentlemen* as substitutes for the person's name. Incidentally, *Gentlemen* has a long tradition of being the correct salutation to use when addressing a group of persons, one or more of whom might be a woman. Many people now consider the salutations *Gentlemen* and *Dear Sirs* (plural) to be outdated.

The standard punctuation to follow a salutation in a North American business letter is the colon. Therefore, many people will regard any other punctuation in that position as an error.

THE BODY OF THE LETTER

Full-block format letters do not contain indented first lines of paragraphs. In fact, the essential characteristic of a full-block letter is that all of its lines begin at a single margin. At the writer's discretion, semi-block format letters may or may not contain indentation. A line space between paragraphs is standard in full-block letters and optional in semi-block letters that have indentation.

THE COMPLIMENTARY CLOSE

Like the salutation, the complimentary close is a formality that is all but invisible — unless it is inappropriate or omitted. It is a graceful way of exiting from the letter. Acceptable complimentary closes are given in the following examples:

Yours truly,	Yours very truly,
Truly yours,	Very truly yours,
Sincerely,	Sincerely yours,
Very sincerely yours,	Yours very sincerely,

Regards,	Best regards,
Cordially yours,	Respectfully yours,
(when the relationship	(when writing
is long-standing and	to persons of
cordial)	high rank)

Note that only the first letter of the first word of a complimentary close is capitalized. Note also that the standard punctuation to follow a complimentary close is the comma. Therefore, many people will regard any other punctuation in that position as an error.

The most important thing to remember when selecting a complimentary close is that it should be, for all practical purposes, invisible. When you get creative in a complimentary close, you run the risk of offending someone, making a fool of yourself, or both. The best course of action is to stick to the standard closes.

THE SIGNATURE GROUP

The signature group appears four or five line-spaces below the complimentary close. It is composed of the writer's typed name and position title. The space between the complimentary close and the signature group is for the writer's signature.

Some firms use a somewhat different method, that of including the company name above the writer's name. This method has some merit, in that it makes clear that the writer is speaking for the company. It is common especially in letters of transmittal.

MISCELLANEOUS OPTIONS AND NOTATIONS

Some letters may require a subject line. In a full-block format, this line normally follows the salutation. In a semi-block format, it is often on the same line with the salutation. The following examples illustrate these two methods of including a subject line:

Dear Ms. Sands:
Subject: Your Request for Proposal 57935

Dear Ms. Sands: Subject: Your Request for Proposal 57935

An acceptable abbreviation is *Re:* in place of the word *Subject.*

If the letter contains enclosures or attachments, a notation of them should appear at least several spaces below the signature group. If copies of the letter will be sent to persons other than the addressee, a notation indicating who these persons are should also be included. When someone other than the writer, such as a secretary or clerk, has typed the letter, it should bear the writer's initials in all capital letters together with the typist's initials in lowercase letters as shown in Sample 9-5. Second and subsequent pages should be handled as shown in the sample letters that follow:

SAMPLE 9-5   A Typical Business Letter in Full-Block Format

 **ellston Corporation**

4567 Airport Road
Sunnyside ON K2V 1M8

Telephone: (613) 555-5000
Fax: (613) 555-5333

An Industry Leader Since 1958

March 5, 1999

Ms. Hortense C. Pembroke
Contracts Administrator
Environment Canada
2356 Garfield Avenue
Ottawa ON K5Y 6M7

Dear Ms. Pembroke:

The Pellston Corporation would appreciate the opportunity to submit proposals in response to future Environment Canada procurements related to weather sensing and signal processing devices. As you know, the Pellston Electronics Division has special capabilities in data gathering and processing, having been a pioneer in geological surveying system development.

In recent years, Pellston's technological base has been expanded to include undersea data gathering systems. As a necessary part of this expansion, we have assembled an outstanding research and development staff, particularly in the field of signal processing. Pellston management believes that this capability places the company in a strong position to expand into yet other fields primarily concerned with data gathering and processing.

The enclosed document, titled *Pellston Capabilities,* is submitted to familiarize you with our recent research and development programs and with our physical facilities.

continued ➜

Ms. Hortense C. Pembroke
March 5, 1999
Page 2

I would be pleased for the opportunity to send any other information that you may require as a qualification for placement of the Pellston Corporation on Environment Canada's list of bidders.

Thank you.

Yours truly,

THE PELLSTON CORPORATION
Fowler D. Moffat
Marketing Manager

Enclosure: Pellston Capabilities
cc: Esther Tant
FDM/esl

SAMPLE 9-6 Letter in Semi-Block Format

Pellston Corporation

4567 Airport Road
Sunnyside ON K2V 1M8

Telephone: (613) 555-5000
Fax: (613) 555-5333

An Industry Leader Since 1958

March 5, 1999

Ms. Hortense C. Pembroke
Contracts Administrator
Environment Canada
2356 Garfield Avenue
Ottawa ON K5Y 6M7

Dear Ms. Pembroke:

The Pellston Corporation would appreciate the opportunity to submit proposals in response to future Environment Canada procurements related to

continued ➤

weather sensing and signal processing devices. As you know, the Pellston Electronics Division has special capabilities in data gathering and processing, having been a pioneer in geological surveying system development.

In recent years, Pellston's technological base has been expanded to include undersea data gathering systems. As a necessary part of this expansion, we have assembled an outstanding research and development staff, particularly in the field of signal processing. Pellston management believes that this capability places the company in a strong position to expand into yet other fields primarily concerned with data gathering and processing.

Ms. Hortense C. Pembroke   Page 2   March 5, 1999

The enclosed document, titled *Pellston Capabilities,* is submitted to familiarize you with our recent research and development programs and with our physical facilities.

I would be pleased for the opportunity to send any other information that you may require as a qualification for placement of the Pellston Corporation on Environment Canada's list of bidders.

Thank you.

Yours truly,

THE PELLSTON CORPORATION
Fowler D. Moffat
Marketing Manager

Enclosure: Pellston Capabilities

Exercises

9-1. Writing a One-Page Memorandum Report

1. Study the following case very carefully.

2. Make an outline for a report that will respond to the situation.

3. Write the first draft of a memo report to Ms. T.S. Chan, Director General of the Alberta Ambulance Authority (address given below), with a copy to Manfred Panfreid, the Vice President in charge of the Management Consulting Division of Pellston.

4. Follow all the steps in the writing process (Chapter 6) in producing a final copy of the memorandum.

THE MEDICAL EVACUATION PROBLEM

Six months ago, Manfred Panfreid, Vice President in charge of Pellston's Management Consulting Division, telephoned you to say that his Division needed your help. All of the consultants were already on assignment, and the Division had received a call from the Alberta Ambulance Authority. They needed someone to immediately take over the coordination of medical evacuation flights throughout the province, and Grant Dobbs told Manfred to give you the assignment. The Alberta Ambulance Authority is located at 4392 Champlain Street, Red Deer, Alberta, T4D 1X8.

Although everything went smoothly at first, you have now encountered a paperwork problem. When emergencies requiring aeromedical evacuations arise, the flights are arranged by the medical authority on the scene. The person arranging these flights is required to complete and submit Form 222B to you within one week after the flight. This form is necessary for the release of funds from Alberta Health Care, 6215 151 Avenue, Edmonton, Alberta, T6B 1C8, for payment of the charter company's fee.

During your tenure as coordinator of medical evacuation flights, 52 flights were carried out, 4 of them by authorization of Dr. Stephen Cooper, Senior Medical Officer of the Harper Hospital in Edmonton. Although many doctors regard Form 222B as a bother, you have had excellent cooperation from everyone but Dr. Cooper.

After the first flight authorized by Dr. Cooper, on July 4, you waited several weeks before telephoning him. He said, rather tersely, that he would get the form to you when he got around to it. On July 29, Dr. Cooper authorized a second flight, once again failing to submit the required document. On August 10, you telephoned Dr. Cooper again, respectfully pointing out that you had yet to receive the documents relating either to this latest flight or the one carried out on July 4. Dr. Cooper stated that he was too busy to discuss the matter, and he hung up the telephone before you could say anything else.

On August 15 and again on August 25, Dr. Cooper authorized emergency medical evacuation flights and failed to submit Form 222B for either of them. On September 20, you telephoned Dr. Cooper once again. During this call, Dr. Cooper not only refused the submit the forms, he also pointed out that he had "better things to do than fill out forms for a clerk (meaning you) who can't get the job done without badgering busy professionals (meaning him)." He also threatened to report you to your superiors and hung up the telephone before you could respond.

The requests you had submitted to Alberta Health Care for payment to the charter companies for the 4 flights have all been rejected because the necessary forms were not included; Archibald Ericson, an official at Alberta Health Care, has hinted that requests for funds unaccompanied by the proper documentation could constitute attempted fraud, a felony; and the general manager of one of the charter companies has telephoned to tell you that she will ignore future requests for emergency medical flights, while going on to explain, expletives included, that she is not in the charity business.

9-2. *Writing a One-Page Letter Report*

1. Referring to the medical evacuation case in Exercise 9-1, make an outline for a letter report to Alberta Health Care, explaining the situation and the action you have taken. You might also recommend changes to their procedure that would help to avoid future problems with people who refuse to fill out forms.

2. Write a report in business letter format to Mr. Archibald Ericson, Coordinator of Medical Evacuation Flights, Alberta Health Care (address given in the case), with a copy to Manfred Panfreid, the Vice President in charge of the Management Consulting Division of Pellston.

3. Follow all the steps in the writing process (Chapter 6) in producing a final copy of the business letter report.

9-3. *Writing a Memorandum Report*

1. Study the following case very carefully.

2. Make an outline for a report that will respond to the situation.

3. Write a first draft.

4. Write the memo report to Grant Dobbs, including a complete description of the problem, a complete discussion of your solution, and the steps you plan to take to remedy the situation. Complete all the steps in the writing process (Chapter 6). Before beginning, carefully review the sample memo report (Sample 9-3) in the chapter.

THE WARRANTY PROBLEM

Mr. Dobbs calls you into his office. He congratulates you on your handling of the ambulance authority matter. This time he has a different sort of a job for you. In recent years, the Pellston Music Division has been less profitable than it should have been. Mr. Dobbs wants you to go out to Surrey and look into the matter for him. You may take as much time as you need to assess the situation, and then do what needs to be done.

Mr. Peabody greets you at the airport. "Well, so you're one o' them whiz kids from the front office, eh?"

"No, sir," you reply respectfully, "I'm just out here to see if there is some way I can help make the operation a little more profitable."

"Well, there ain't no puzzle there, kid. It's our warranty program. It's killin' us slowly but surely — repairs on our guitars especially. We have to quit givin' that five-year warranty on parts and workmanship."

During the drive from the airport, Peabody elaborates on the problem. He believes that most of the guitars that are returned for warranty work have not been properly cared for. He adds that it is impossible for the factory to prove that they have been abused because the dealers handle all the warranty work on a local basis. He also believes that customers might lose confidence if the company tried to explain that rapid changes in temperature or humidity can cause problems that are unavoidable even in the most expensive instruments. Peabody states further that many people who travel with instruments don't take sufficient care of them when they are in transit.

After you arrive at the plant, Mr. Peabody tells you to feel free to talk to anybody about anything. Your first stop is accounting, where you discover that Peabody was absolutely right. The profit margin as a whole is reduced drastically by the cost of fulfilling the warranty program. Abigail Barash, a senior accountant, explains that all warranty work is done by dealers' shops, whose hourly labor charges run as high as $50. Last year alone, the company paid out over $436,000 just for labor.

For the next week, you painstakingly go over all of the charges for warranty work (done on guitars alone) by this dealer network. You find that many of the dealers send out the work to local repair shops and then tack on an additional charge for their handling of the instrument. You also discover that of the total repair work, 48 percent was required because of improper storage, 37 percent was done to repair instruments that were damaged during travel, and another 7 percent was probably the result of flagrant abuse. Therefore, only 8 percent of the expense was justified. And if the repairs on that 8 percent had been done at the factory, the actual cost would have been $19,512, instead of the $34,880 the company paid to the dealers.

You decide that, even if Pellston bears the cost of shipping, the company can save over $400,000 a year by doing all warranty work at the factory. The warranty clearly states that damage resulting from improper storage and handling is not covered in the warranty. This policy needs to be enforced assiduously. But you are worried about the effect on sales that such strict adherence to policy might bring about. Are all of these people who return damaged instruments just trying to get something for nothing? Or do they simply not understand that musical instruments made of wood need special care and attention?

You have another, even more careful, look at the records, and you discover that nearly 25 percent of the repairs have to be done more than once, adding still more to the total cost. This fact suggests that these repairs are probably being done by persons without proper training and equipment. And this latter fact suggests a method of selling the idea of factory repairs to customers, because changing the terms of the warranty to assure better repair work is a benefit to the customer.

You dig out a copy of the present warranty booklet, which has a cover that simply says, "Congratulations!" Inside, the booklet says:

> You have just purchased a Pellston guitar, which has more than 80 years of tradition built into it. We believe that no guitar of similar price on the market today can match the quality of materials and workmanship of a Pellston, and for that reason we are pleased to offer the following warranty.

Warranty

> Pellston Music Division warrants this guitar from all defects in materials and workmanship for five full years after date of purchase. Pellston will not be responsible for normal wear or for any damage caused by improper handling or storage. Also, Pellston assumes no liability beyond the restoration of an instrument that has failed due to faulty materials and workmanship.

Your next stop is the factory floor, where Roy Bullfinch, who has been with the company for almost as long as Mr. Peabody, is happy to tell you all about guitars and the way to keep them from being damaged. You write down the following points:

1. Always put the guitar in its case when you aren't playing it. Leaning it against the wall just invites damage.

2. In a damp climate, a couple of desiccant bags in the case will keep the humidity down. Otherwise, the wood might swell, and the glue joints might come loose as well.

3. In a dry climate, one of those little containers that has holes in the top and that has a moistened sponge inside can be put in the case. These containers are available in almost any music store.

4. Temperature extremes are bad for the guitar. Cold temperatures will cause cracks in the lacquer finish, and these cracks are a sure sign of improper storage. The trunk of a car, where temperature extremes are common in practically any climate, is no place for a guitar.

5. Generally speaking both hot-and-humid and cold-and-dry extremes will most likely damage a guitar.

6. Strings should be loosened before storing for long periods of time (six months or more) or before traveling by car or public transit for long distances.

7. Strings should not be loosened and tightened more often than necessary. It doesn't do the guitar any good, and it ruins the strings.

8. The guitar should be kept free of fingerprints. The oil in your skin will eventually ruin the finish.

9. Some furniture polishes are harmful to a guitar finish. Some of them contain silicones or other compounds that ruin the finish. Guitar polish is available at many music shops.

10. Straps made of artificial materials, such as vinyl, contain solvents that will mar the finish of the guitar. Therefore, the best course of action is to use only cloth or leather straps.

11. Pellston's steel-string guitars have sealed tuning machine mechanisms that do not require oiling. The classic models have open mechanisms that need light lubrication. This operation should be done every six months or so, and care should be taken to ensure that the mechanisms are kept clean.

12. Some bulging around the bridge area can occur if strings heavier than those recommended for a specific model are used. If such use is continued, the bridge can pull off.

You also check with several airline companies to find out what their policy is on guitars that might be damaged in transit. All of them tell you the same thing. They will pay up to $500 for repairs if the guitar has been damaged by improper handling by their personnel. However, for a small fee, you can insure any piece of baggage for a larger amount. Moreover, you can bypass the conveyor system, where most damage occurs, by asking for special handling by the airline.

Next, you talk to the people in shipping to find out how they ship completed guitars from the factory to the warehouses. You discover that they have special shipping cases for each model, and that a supply of these cases could be kept at the warehouses for return of instruments to be repaired under warranty.

You now have all the information you need to go ahead. Putting your ideas into operation will require four communications: a memo report to Mr. Dobbs with a copy to Mr. Peabody; a letter to all dealers regarding warranty work; a set of directions to warehouse personnel, telling them about their new responsibilities for shipping and receiving warranty work; and a rewritten warranty booklet that will include instructions for caring for the instrument.

9-4. *Writing a Letter Report*

1. Study the following case very carefully.

2. Make an outline for a letter report that will respond to the situation.

3. Write the report in letter format. The addressee will be Mr. Juan Martinez, President, South American Enterprises, 9000 Avenida Del Bar, Santiago, Chile.

4. After reviewing the material on letter format and Samples 9-5 and 9-6, write the first draft.

5. Rewrite the draft, following the steps given in Chapter 6.

Some Special Considerations:

- Be very wary of simply repeating the terminology used in the case. "International conglomerate," for example, is often used as a term of derision. And slang terms, such as "hard nose" and "bean counters," don't belong in a technical report either.

- You will need to reorganize and rewrite all the information in this case. Resist the temptation to "lift" sentences and expressions. Remember that the information in the case is composed of notes, randomly thrown together. Also resist the temptation to narrate events or to quote those who were interviewed.

- Although this report is to be done in letter format, bear in mind that, as a report, it may contain headings. You would be wise to begin with an abstract. And don't forget the "subject" line.

THE SPACE PRODUCTS, INC., PROBLEM

The Pellston Management Consulting Division has been called in to assess a serious drop in productivity at a company called Space Products, Inc. (SPI), located at 1733 Market Boulevard, St. John, New Brunswick, B8H 9B6. Ralph Kaplan, one of Pellston's business consultants, has completed his study of the problem. Unfortunately, the technical writer in the Management Consulting Division, Sally Green, has come down with a severe case of flu. South American Enterprises, the parent company of Space Products, expects a report right away; therefore, Mr. Panfreid hands you Ralph's notes, saying, "Turn this into a report right away. SPI is losing money every day, and South American Enterprises wants our findings on this as soon as possible."

RALPH KAPLAN'S NOTES

SPI production is down 18 percent overall. The company is losing about $30,000 a day.

Observed causes: Design mistakes have made extensive redesign necessary during production engineering stages on the last two development projects. Employees working on the assembly lines are openly hostile to management. Work slowdowns have occurred, not as a result of any orchestrated effort on the part of the union, but as an apparently spontaneous response to management directives ordering increased production.

Background information: Management completely changed eight months ago when the company was purchased by an international conglomerate. The new owners replaced all of the key people, hiring hard-nose types.

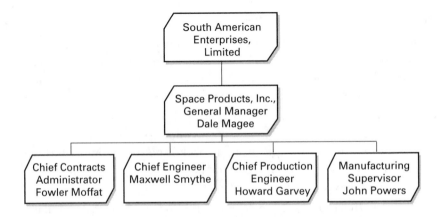

Organization chart borrowed from an official SPI publication

They got several of these people from a consulting firm known as Efficiency Expertise, Inc. At Efficiency Exp., Dale Magee had headed up a department responsible for assessing efficiency at client factories. When South American hired him, he hired Smythe, Moffat, Garvey, and Powers away from his old company.

Before the change in ownership, SPI was noted both for efficiency and high-quality products. When the new owners replaced the existing regime with new people, all of the employees, from the project engineers to the workers on the factory floor, were deeply resentful. All the employees had had an excellent working relationship with the old management group. Some had worked for the same boss for more than 20 years.

Some of the changes made since the new management team took over are as follows. Punch clocks have been installed in the Engineering Department. Most of the engineers had been accustomed to working 50- and 60-hour weeks. As professionals, they never expected to receive (nor did they ever receive) overtime pay. The purpose of the clock is not to compensate them for their overtime work, but merely to keep track of who is working long hours and who isn't. Now that they have to clock in and out, most don't bother to work overtime. They feel as though they are being pitted against one another. One engineer confided that he doesn't feel as though the management and the engineers have the same goal any longer. He said, "Before, everyone worked hard to get the job done and to get it done right. Now, the objective is to get through the day. They (the managers) are on one side, and we're on the other."

Mr. Magee (himself) has timed many of the secretaries to see how long it takes each one to type a letter. Five of these secretaries have quit as a result. One had been with the company for 23 years. Morale among all of the employees could not be lower. When I asked the question: "How do you think your boss regards you?" one of the secretaries used the term "galley slave." Another's answer was, "not very highly — and not for much longer either." She quit before I completed my study, the fifth secretary to quit in the past six months.

Supervisors in the plant have been instructed to enforce 15 percent higher production quotas for each position. Several of the previous supervisors have quit as a result, and they have been replaced. John Powers insists that the quotas are realistic, although production has been 15 percent to 20 percent below the original quotas since the new quotas were established. On paper, Powers appears to be right. The higher quotas are numerically possible. On the factory floor, Powers is dead wrong. The machines may be capable of producing the new quotas, but the machine operators are not. Powers has failed to take into account the delicate work of setting up for each run. The operators can't be rushed in this part of the job.

I conducted interviews in the plant. One term I heard used often was "bean counters," often with an expletive before it. The workers in the plant regard management as the enemy. They take pride in doing as little work as possible. Some workers have gone so far as to intentionally damage machines to avoid the impossible quotas. Others put out the quotas with time to spare, but as much as 50 percent of the run gets rejected when the parts get to the Inspection Department. Inspectors, too, are made to do the job faster. As a result, out-of-spec parts have gone into assemblies, and one government contract was terminated as a result of the poor reliability of the product.

The community at large has a low opinion of SPI. Literally, nobody in St. John has anything good to say about the company. Everyone I talked to agreed that the union and the company were headed for a showdown. The union contract will be up in two months, and all the employees I talked to said that they were prepared for a long strike. Management says the strike won't bother them; they will simply bus in non-union workers and keep the plant running. "Anyone who doesn't want to work here doesn't have to," is the way Dale Magee put it. He said also that anyone who goes out on strike will never work for SPI again. He says they will be fired automatically.

No deep mystery here. One need only consider that company profits in the 12-month period before the change of ownership were nearly $30 million, as opposed to the projected loss for this year of nearly $8 million. The only difference is in the style of management. The current management philosophy in this company is right out of Charles Dickens. It is something of a shock to find that a company that has operated successfully for 25 years can be nearly destroyed in a matter of months by nothing more than archaic management ideas.

The cure is to get rid of Magee and all of the other department heads he has brought in. The recently hired staff at lower levels may be okay under a different manager. A new general manager must be appointed before the union negotiations begin next month. Discreet inquiries should be made to find out if Jean Polk would be interested in having her old job back as general manager.

Magee has a two-year contract, but the damage he could cause in another 16 months would likely be fatal to the company. His contract should therefore be paid out without delay. Although this buy-out will cost the company $135,000, it will be a small price to pay when compared to the $150,000 being lost each week under the current regime.

If Polk isn't interested (I understand she has gone into early retirement), someone else must be found who has a knowledge of modern management techniques. The company should be able to regain most of its strength under an enlightened management.

Special note to Sally Green: South American uses private employment agents extensively in this country for the simple reason that their corporate structure is strictly a financial one. They don't have a person to send here to hire management or to troubleshoot a problem. That is why they hired us. The employment service that gave them Magee and his group should not be used again. Anyone who couldn't figure out Magee in a three-minute interview shouldn't be in the personnel business. We will have to soft pedal all of this, I suppose. Make it strong but impersonal.

9-5. *Writing a Memorandum Report That Includes a Heading Structure and Illustrations*

Although the information in the case below appears quite technical at first glance, *no technical knowledge is required to complete this exercise.* You will find that with a little effort you will be able to understand the situation quite easily. Thus, you will have the opportunity to relate technical details to support a rather simple logical framework. Although this is a memorandum report, its content should reflect an appreciation of the importance of full disclosure. Use the diagrams given in the case.

1. Thoroughly review Sample 9-4 in the chapter before beginning this assignment.

2. Study the following case very carefully.

3. Make an outline for a report that will respond to the situation. Here is a possible working outline to get you started. You may use the outline as it is or add to it (or change it completely) as you progress.

Abstract

I. Specification

II. Design Details
    A. Case
    B. Battery
    C. Circuit Performance
        1. Transmitter
        2. Receiver

III. Environmental Testing
    A. Procedure
    B. Results

IV. Reliability

V. Conclusion

4. Write a first draft of the report; write the introduction and conclusion last.

5. Use the illustrations given in the case. Assign a figure number for each illustration and refer to each illustration by its figure number before placing it in the text. Also, supply a caption, comprising the figure number and a suitable title, for each illustration.

6. Place data that lends itself to tabular form into tables. Handle the tables in the same way as you did the illustrations (reference by number, placement of the table after the reference, and a caption composed of a table number and a suitable title), but give the tables a separate sequence of numbers.

7. Write a final draft of the report, following all the steps given in Chapter 6.

THE POLICE TRANSCEIVER REPORT

You have been given the job title of Staff Assistant and have been assigned to Darla Petoskey, Chief Engineer of the Electronics Division. She has called you into her office today to talk about the Firefly Transceiver.

"We are trying to improve the Firefly by redesigning to a new specification," she says. "The police in several large cities have found it to be unreliable at times and a bit too bulky for easy use."

"Is the Firefly the sort of radio that you see SWAT teams using on those television police shows?" you ask.

"Yes, exactly," Darla replies. "In response to some of the suggestions we've received from the various police forces using these radios, we've come up with a detailed specification for the Super Firefly. The company allotted funds for one of my engineering groups to design and test a prototype. Now that we are almost finished, the next step is for the Production Engineering Department to take over the project and modify the design so that it can be mass produced.

"We need to put out a report for corporate management. They will have to review the project before putting an okay on the money for production engineering and tooling."

"And that's where I come in?"

"Right," Darla continues, "and since the report is an internal one, it can be in a memorandum format. However, it has to contain all the information on the project, so that the brass upstairs can make a proper decision. Here is a copy of Specification PEL 31-33. This specification covers all of the performance requirements and space limitations of the unit."

"I'll study this later," you tell her while listening for more details.

"You'll find that we've done pretty well in meeting the requirements of the spec. We devoted most of our attention to transmitter performance, since the transmitter was somewhat unreliable in the original Firefly and since we don't see any problem with the present receiver section. Also, the overall size constraints have been met.

"Weight computations have been made. I'll give you a copy of those later, along with some design details of the case and cover. Say, why don't I just turn over my daily log book to you. I'll get some duplicate copies of the relevant pages of Joan Santiago's book too. She's the Project Engineer, and she has done most of the work herself."

Here is the specification Darla mentioned.

SPECIFICATIONF PEL 31-33

1. Dimensions and Weight Limitations

    1.1. The overall dimensions of the case shall not exceed 15.24 cm by 10.16 cm by 5.08 cm.

    1.2. The overall weight shall not exceed 340.2 g.

2. Power Output and Transmit Duty Cycle

    2.1. The power output shall be 1 watt minimum, and the transmitter shall be capable of producing this power at temperature extremes of –40 degrees C and +51.7 degrees C.

    2.2. The design shall include batteries capable of providing power to the unit for 8 hours with a 10 percent transmit cycle, and 4 hours with a 30 percent transmit cycle.

3. Mechanical Design

    3.1. Mechanical packaging shall provide assurance of watertightness, and all exterior surfaces shall be suitably protected with waterproof finishes.

4. Mean Time Between Failure (MTBF)

    4.1. The MTBF shall exceed 5000 hours.

5. Environmental Testing

    5.1 The unit shall demonstrate operational capability as specified in temperature extremes of –40 degrees and +51.7 degrees C.

    5.2 The unit shall operate as specified after receiving a shock of 12 g's in any of the directions perpendicular to its surfaces and points joining them.

    5.3 The unit shall operate as specified after vibration from 10 to 2000 cps at a double amplitude of 0.508 cm, in steps of 10 cps (from 10 cps to 100 cps) and in steps of 100 cps (from 100 cps to 2000 cps). In addition, vibration frequencies shall be cycled to find all resonant frequencies in the specified range, and the equipment shall be subjected to each of these in turn for 1 minute.

    5.4 The unit shall operate satisfactorily after having been submersed in water at a depth of 3.048 meters for 30 minutes.

6. Transmitter Efficiency

    6.1 The unit transmitter efficiency shall exceed 40 percent.

[end of specification]

Here is the data from the logbook of Joan Santiago:

Transmitter output stage efficiency

        dc input 205 ma $\times$ 8 v = 1.639 watts

        RF out = 1.120 watts

        RF in = 0.2 watts

        Efficiency = 1.120 – 0.2 = 0.920/1.639 = 56.25%

        Oscillator stability at 45 degrees = 0.00062%/hr

Transmitter current demand

> oscillator stage = 2.5 ma
>
> buffer stage = 8.0 ma
>
> first doubler stage = 18.0 ma
>
> driver amp = 70 ma
>
> second doubler stage = 303.5 ma

Measured RF voltage

> 7.5 vrms. Computed power out = $(7.5)2 = 1.12$ watts

Notes: Careful impedance matching has yielded an efficiency significantly greater than the design parameter of 40 percent. Some efficiency improvement will be lost when the transmitter is packaged into the unit as a whole. However, the margin of improvement is great enough to compensate for the loss. After cold tune-up, the frequency was adjusted to 550 Hz from the nominal value. The SK-10-board circuit was then operated for 1 hour at ambient temperatures (26 degrees C at the beginning to 30 degrees C at the end of 45 minutes). The results are max. change of –500 Hz at 8 minutes to +250 Hz at 54 minutes.

Notes from the logbook of Darla Petoskey:

Mechanical Packaging

The environmental test specs (paragraph 5) dictate some method of encapsulation. Supported rigid urethane foam chosen.

The mechanical design meets all of the criteria set forth for it. Here is a drawing of the prototype.

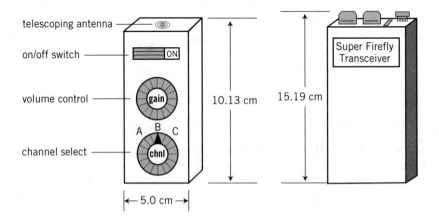

All space, weight, and reliability requirements have been met. Dimensions are shown on the drawing above. Overall weight = 327.44 g, calculated as follows:

Case	=	94.12 g
Circuit boards	=	59.53 g
Components and solder (on prototype)*	=	84.48 g
Controls and antenna	=	89.31 g

   ***Note**: Solder weight will likely decrease on production models.

Battery Details

Power to be supplied by two specially built 6.5-volt lithium batteries. Batteries to be furnished by Maxivolt Battery Co. of Winnipeg, Manitoba. Batteries fit into the back of the unit as shown in the following diagram.

Two specially constructed
6.5 volt batteries

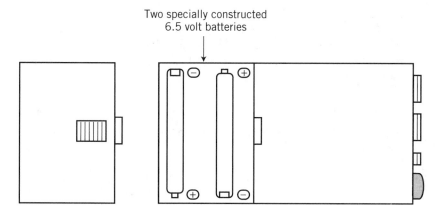

Results of Tests in Simulated Environment

The unit was operated as specified in temperature extremes of –40 degrees C and +51.7 degrees C. Measured power output was 1.05 Watts at both extremes, reaching 1.12 Watts at 20 degrees C.

The unit was mounted on the shock table and subjected to a shock of 12 g's in all directions perpendicular to its surfaces and perpendicular to points joining them. Operation was normal following this procedure.

The unit was mounted on a vibration table and subjected to vibration from 10 to 2000 cps at a double amplitude of 0.508 cm., in steps of 10 cps (from 10 cps to 100 cps) and in steps of 100 cps (from 100 cps to 2000 cps). Vibration frequencies were then cycled to find all resonant frequencies in the specified range; the equipment was subjected to each of these in turn for 1 minute. Following the first attempt, the unit would not operate. The problem was found to be loose solder connections in both the transmitter and receiver sections. The loose components were remounted and resoldered. The unit was then encapsulated in No. 714 Potting Compound (supported rigid urethane foam). The vibration procedure was repeated. Following the second vibration procedure, the unit operated satisfactorily.

The unit was submersed in water to a depth of 3.048 m for 30 minutes. Following submersion, operation was normal.

Transmitter Efficiency

The unit transmitter efficiency exceeded the specified figure of 40 percent.

Battery Operation

With a battery pack installed, the unit operated for 9.1 hours when the transmitter was operated on a 10 percent duty cycle. The unit operated for 4.8 hours on a transmitter duty cycle of 30 percent.

Reliability Forecast

Reliability calculations indicate a mean time between failure (MTBF) of 8240 hours.

CHAPTER 10

Long Reports

OBJECTIVES: In this chapter, you will learn about reports that are required as line items of contracts and reports that are submitted for other reasons, such as a problem with a product. You will also learn that most long contractual reports are written in accordance with a specification, such as the one given as a sample in Chapter 7.

The Elements of a Long Report

A long, or full-scale, report includes front matter, the report proper, and appendices as needed. The report proper, occasionally called the report body, has a heading structure, the first heading of which is often *INTRODUCTION*. Traditionally, the introduction is not considered one of the front matter elements, which are discussed below.

Front Matter

The front matter of a full-scale report contains some or all of the following elements: a cover, letter of transmittal, title page, abstract, table of contents, list of illustrations, and list of tables. Each of these items is discussed separately in the following paragraphs.

The Cover

External reports are often bound inside a standard cover. Some companies have a stock of covers that have a die-cut opening through which information from the title page can be seen. The advantage of the die-cut cover is that the title page serves a dual purpose: the title and date show through the cut-out area and thus form part of the cover.

If die-cut covers are not used, the title and date of the document can be lithographed or typed directly on the cover stock or on labels, which can then be placed on each cover as shown in Sample 10-1.

The Letter of Transmittal

When you send someone a gift, you probably include a card with some expression of best wishes on the occasion that prompted the gift. A letter of transmittal serves a purpose similar to the card that accompanies a gift; that is, it tells the person exactly why the document is being submitted. It may also say something about the document itself, and it may acknowledge persons outside the writing team who contributed in some way. The letter of transmittal is also the proper vehicle for citing any exceptions taken to specifications or other requirements.

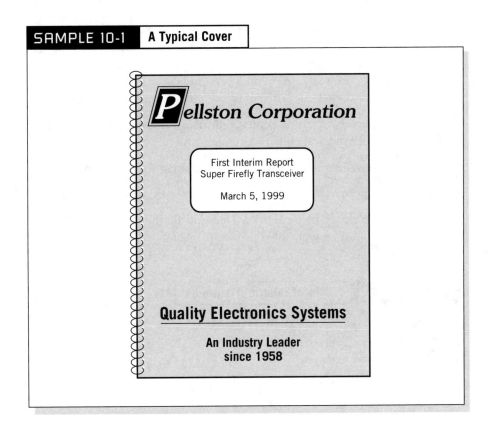

A writer of a report may occasionally write, but almost never signs, the letter of transmittal. More often than not, letters of transmittal are written and signed by a contracts administrator, a marketing representative, or some other company official. There are two reasons for this practice. First, the letter should be written and signed by the sender of the document, who is not very often its originator. Second, and far more important, letters signed by persons other than company officials can open unauthorized channels of communication. (Refer also to the following discussion of the title page.)

Some firms, regarding the letter of transmittal as an integral part of a document, bind a copy of the letter inside each copy of the document. This practice changes the character of the letter of transmittal, making it more of a preface statement than a true letter of transmittal. Often, firms that subscribe to this practice find that a second letter of transmittal is necessary to fulfill the transmittal function.

It should be noted also that a separate letter of transmittal should be prepared for each recipient. In other words, the copies sent to the customer will require a letter of transmittal different from those that convey copies to the National Library or to company field offices.

Normally, only one letter of transmittal will accompany each package of documents submitted. Therefore, if multiple copies of a document are sent (the rule and not the exception), one letter of transmittal is packaged with the shipment or is, occasionally, sent under separate cover.

As already mentioned, the letter of transmittal may do four things:

- Convey the document, citing contract numbers or other information pertinent to the origins of the document.

- Provide some salient information about the document's contents.

- Indicate exceptions taken to specifications or other stated requirements.

- Express gratitude for help received from outside the report team. Sample 10-2 shows a typical letter of transmittal.

SAMPLE 10-2 | **A Typical Letter of Transmittal**

Pellston Corporation

4567 Airport Road
Sunnyside ON K2V 1M8

Telephone: (613) 555-5000
Fax: (613) 555-5333

An Industry Leader Since 1958

February 21, 1999

Ms. Hortense C. Pembroke
Contracts Administrator
Environment Canada
2356 Garfield Avenue
Ottawa ON K2S 4V9

Dear Ms. Pembroke:

Six copies of the report titled "Final Report, Feasibility of Automatic Remote Weather Stations" are herewith submitted in compliance with Line Item 7 of Contract WB-88-0076. As stated in the report, our study shows that such a system is not only feasible, but that it can be built for a sum significantly lower than the specified target figure.

No exceptions are taken to the performance specification included as part of the guidelines for the system.

continued ➤

We express our sincere thanks to Professor Murray Grey of McGill University for his generosity in allowing us to test several concepts in his meteorological laboratory.

We would be pleased to answer any questions that may arise in connection with the study.

Yours truly,

Barton Vishinsky

THE PELLSTON CORPORATION
Barton Vishinsky
Contracts Administrator

Enclosure: Six copies of "Final Report, Feasibility of Automatic Remote Weather Stations"

The Title Page

The title page should contain the following entries:

- The document title in full
- The date
- The words *Submitted to* followed by the full name of the customer's company
- The words *Submitted by* followed by the name of the company submitting the document
- The contract number, if applicable
- Other items as directed by governing documents

The name of the author(s) should *not* appear on the title page. Companies have an excellent reason for omitting names of authors from title pages. Including authors' names can seem like an invitation to customers to contact the authors directly, and such unofficial channels of communication can cause difficulty. Customers are normally restricted to communication through marketing and contracts people, who are officially charged with such responsibilities. When customers start telephoning a company's engineers and technicians to get information or make suggestions, a company has lost control of its operations.

The title page information should be balanced on the page, as shown in Sample 10-3.

FINAL REPORT
FEASIBILITY OF REMOTE
AUTOMATIC WEATHER STATIONS
(Contract WB–88–0076)

February 21, 1999

Submitted to:

Environment Canada
Ottawa, Ontario

Submitted by:

THE PELLSTON CORPORATION
Sunnyside, Ontario

Abstracts, Executive Summaries, Summaries, and Summary Digests

The "top-down" organizational pattern, which good technical writers use as an overall scheme in all the technical reports they write (Chapter 6), includes a disclosure of all significant facts at the very beginning of the document. To fulfill this purpose, writers briefly summarize the contents somewhere near the front of the document, usually immediately following the title page. If the document is from 10 to 100 pages long, this preliminary information is usually called an *abstract* or an *executive summary*. If the document is longer than 100 pages, this information is often called a *summary*, even though some people object to the use of this word to describe an element coming at the beginning of a document.

A very long document, such as a multi-volume report, may also contain a *summary digest.* For example, a six-volume report on the feasibility of a scientific observatory on the lunar surface included a 90-page summary and a 31-page summary digest.

The abstract in a technical report is not like an abstract that would appear in a book of abstracts. The purpose of abstracts contained in books of abstracts is to tell readers what the documents are about. They are intended to give the reader enough information to decide which documents are related to a specific area of interest. Such abstracts, sometimes called *descriptive abstracts,* are inappropriate for use in technical reports because they do not summarize the report.

Following is an example of a descriptive abstract, which one might find in a book of abstracts.

ABSTRACT

This report describes a feasibility study for an automatic remote weather station. The study was carried out by the Pellston Corporation under contract to Environment Canada. Areas covered are data-gathering instruments, housing requirements, telemetering systems, and maintenance scheduling.

An abstract in a technical report briefly summarizes the contents of the report, giving particulars of findings and recommendations. These abstracts, sometimes called *informative abstracts,* are appropriate in technical reports. An example of an informative abstract is given below.

ABSTRACT

This report describes a feasibility study for an automatic remote weather station. The study shows that such weather stations are feasible and that no special instrumentation would need to be developed, provided that the housing structure is adequately designed. Of the several housing structures considered, a welded steel-alloy framework with replaceable fiberglass panels was found to be the best and cheapest.

Data would be telemetered through the existing Delta-5 Satellite. The data can be received and recorded on equipment already in place at Central Weather Advisory Centers.

The maintenance schedule will consist of a quarterly routine that will take one worker-day per installation.

The abstract, executive summary, or summary usually appears by itself beginning on the page immediately following the title page.

You will find that writing the abstract is an easy task if you follow six simple steps. Refer to the preceding example as you read through these steps.

1. Do not attempt to write an abstract until *after* you and your technical writing group members have written everything else.
2. Write one or two sentences that tell what the document is about.
3. Write one or two sentences that state the document's conclusions.
4. Reviewing your own heading structure, write a sentence describing what information is contained under each primary head. Use the topic sentences in key paragraphs in the document for this purpose.
5. Combine all of the sentences you have written.
6. Rewrite and edit as necessary, following the steps in the writing process given in Chapter 6.

The Table of Contents

The table of contents lists all headings contained in the report and gives the respective page number for each. In very long reports (200 pages or more), writers sometimes find it necessary to list only the first two or three levels of headings, in order to avoid extremely long tables of contents.

The table of contents should be labeled as such, and it should provide column headings as shown in Sample 10-4. Under no circumstances should the word *page* be repeated before each page number.

SAMPLE 10-4 | **Table of Contents**

TABLE OF CONTENTS

Section	Item	Page
	Abstract	2
	List of Illustrations	4
	List of Tables	5
I	Introduction	6
II	System Description	7
	A. Instrumentation	7
	B. Housing Considerations	16
III	Maintenance Requirements	28
	A. Power Generation Systems	28
	B. Instrument Standards	36
IV	Conclusion	43
Appendix A	Concept Test Results	45
Appendix B	Instrument Specifications	53

Note that the table of contents includes the abstract, even though most formats require the abstract to come before the table of contents. Note also that the table of contents does not include the title page, the letter of transmittal, or the table of contents itself.

The List of Illustrations

Every illustration will have a figure number and title, both of which must be listed in the list of illustrations, along with the number of the page on which the illustration appears. The list of illustrations is usually placed on the page following the last page of the table of contents. Sample 10-5 shows a typical list of illustrations. Note that three column headings are used. Under no circumstances should the word *figure* or the word *page* be repeated for each entry.

SAMPLE 10-5 | **List of Illustrations**

LIST OF ILLUSTRATIONS

Figure	Title	Page
1	Instrument Placement	8
2	Telemetering System Block Diagram	12
3	Construction Technique	18
4	Power Generation System	32

The List of Tables

The list of tables is similar in format to the list of illustrations. Tables are assigned a sequence of numbers separate from the sequence of figure numbers, and they are therefore listed separately. The list of tables usually appears by itself on the page immediately following the last page of the list of illustrations. However, if both the list of illustrations and the list of tables are short enough to fit on a page together, such an arrangement is permitted in most formats. Under no circumstances, however, should they be joined within a single sequence of numbers. Sample 10-6 shows a typical list of tables.

LIST OF TABLES

Table	Title	Page
1	Instruments and their Manufacturers	9
2	Performance Criteria	14
3	Materials Costs	19
4	Maintenance Routine	35

The Report Proper

The report proper usually begins with an introduction, the purpose of which is to provide background information and smooth transition into the technical material that follows. The report proper often ends with a conclusion, recommendation, or both.

The Introduction

Students often become confused when writing introductions. The confusion usually arises from the similarities between summaries and introductions. Introductions do not summarize the contents of a document. An introduction gives background information that prepares the reader to understand what is to follow. Occasionally, you will make statements that are essentially the same in the summary or abstract and in the introduction. In fact, you may already have said something similar in the letter of transmittal. This kind of repetition is normal, and you need not be uncomfortable with it.

A sentence in the letter of transmittal might say, for example:

Pellston's proposed re-work of Radio Set AN/ART-33 has produced a device that meets or exceeds all the design goals and performance requirements set out in Specifications CAN-R-9898G and CAN-P-7564-TS.

A sentence somewhere in the abstract of the same report might read as follows:

> The prototype test program carried out as part of the development program for the AN/ART-33 Radio Set has shown that all design goals have been met or exceeded, and all paragraphs of Specifications CAN-R-9898G and CAN-P-7564-TS have been met.

And the introduction might begin as follows:

> The purpose of this report is to provide details of Pellston's development program to redesign Radio Set AN/ART-33 in accordance with the goals set forth in Specifications CAN-R-9898G and CAN-P-7564-TS.

Such necessary repetition is not obtrusive. Each of the elements, the letter of transmittal, the abstract, and the introduction, serve different purposes, and their essential structures are also different.

The introduction can follow any of a number of patterns. It can begin with a statement of the purpose of the report or the purpose of the project reported on. The only caution here is to make certain the two are not confused. One way to avoid confusion is to place subheadings in the introduction.

INTRODUCTION
> Purpose of report
>
> Purpose of project
>
> Scope of report
>
> Plan of report development

The introduction should mention any background information that would help the reader to understand what is to follow. How did the project come about? Why was it considered necessary? Sample 10-7 shows a possible opening for an introduction.

SAMPLE 10-7   **A Report Introduction**

INTRODUCTION

Many downed pilots could be saved if a light, automatic, affordable electronic locator transmitter were available on the market. The purpose of the AN/ART-33 Radio Set development program was to design such a device. This report provides details of the design and test program ...

The scope and limitations of the report should also be given in the introduction. What aspects of the development are included? What aspects are excluded and why?

The final part of the introduction can offer a brief description of the plan of development of the report. In effect, this part of the introduction is a brief restatement of the table of contents, but in very broad terms.

Writing Style

The writing style appropriate for a long technical report is similar to that of the letter report. Both letter reports and long reports must be written so that anyone interested will be able to understand them. However, letter reports do occasionally include names and pronouns. In general, long reports omit personal pronouns and names of persons. Consider the following examples:

Letter Report Style

As you requested, I went over the test procedures, and I did indeed discover an error that could have produced the anomalies in the results.

Long Report Style

A review of the test procedures revealed an error that could have produced the anomalies in the results.

The difference apparent between the foregoing examples is the phrasing to omit the personal pronouns. Such phrasing does not, as some people would argue, make long report style cold and impersonal. Long report style focuses on facts, while at the same time eliminating any tendency to narrate events.

This tendency to narrate must be curbed even in letter reports. The narrative approach, which requires a chronological ordering of events, can become extremely cumbersome. Although certain kinds of information demand presentation in chronological order for logic's sake, the narrative style of writing is inappropriate in nearly all technical reports.

Page Layout

Pages with narrow margins and long paragraphs discourage readers before they even begin. Short paragraphs and generous margins on all four sides are, therefore, desirable.

Readers are even more discouraged when they find an illustration that hasn't been referred to in the text, or when they find a reference to an illustration and can't find the illustration. Knowledge of two of the most fundamental principles of sound page layout will help you to avoid these errors. Here are the two principles:

- The text (linear mode of exposition) must carry the burden of continuity.
- Illustrations (spatial mode of exposition) and tables (parallel mode of exposition) must appear as soon as possible after their references.

The first principle means that illustrations and tables cannot just be dumped into the text. *These devices are for the purpose of augmenting, not replacing, text.* Naturally, an illustration is a *de facto* replacement for text, since its inclusion may make long and elaborate explanations unnecessary. However, the text must refer the reader to the illustration, since the two modes of expression are essentially different. Likewise, tables differ from text in that they present information in parallel rather than linear, syntactical forms.

Appendices

Related but peripheral information, such as derivations of formulas or other supporting data, should be placed in appendices. Such an arrangement precludes the likelihood of a cluttered text, since it allows for the inclusion of all pertinent information without the requirement for its placement in the report proper.

The following example of a technical report gives an idea of how a report should look:

Sample cover with die-cut for title

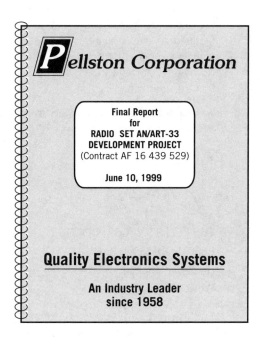

Sample Letter of Transmitttal

 **P**ellston Corporation

4567 Airport Road
Sunnyside ON K2V 1M8

Telephone: (613) 555-5000
Fax: (613) 555-5333

An Industry Leader Since 1958

June 10, 1999

Dr. Joan B. Gonzales
Contracts Administrator
Aviation Equipment, Ltd.
5567 Armbruster Avenue
Halifax NS B8H 9Y9

Dear Dr. Gonzales:

Eight copies of the final report on the Radio Set AN/ART-33 Development Project are herewith submitted as Line Item 7 of Aviation Equipment, Ltd., Contract AF 16 439 529. All the design goals have been met or exceeded, and all paragraphs of Specifications CAN-R-9898G and CAN-P-7564-TS have been met.

We thank you for the opportunity to develop the AN/ART-33, the manufactured version of which is to be designated AN/ART-34. Preproduction models are currently being fabricated and will be forwarded for test on November 1, along with the appropriate test procedures, per Items 8 and 9 of the contract.

We would be pleased to answer any questions that may have been unanticipated in our final report.

Yours truly,

Harold T. Mitchell

THE PELLSTON CORPORATION
ELECTRONICS DIVISION
Harold T. Mitchell
Project Engineer

Enclosure: Eight copies of "Final Report for Radio Set AN/ART-33
            Development Project"

Page 1 of 129 pages

FINAL REPORT

ON

RADIO SET AN/ART-33

DEVELOPMENT PROJECT

(Contract AF 16 439 529)

June 10, 1999

Submitted to:

Aviation Equipment, Ltd.

Halifax, Nova Scotia

Submitted by:

PELLSTON ELECTRONICS

DIVISION OF

THE PELLSTON CORPORATION

Sunnyside, Ontario

ABSTRACT

Radio Set AN/ART-33 is a solid-state emergency locator transmitter with a two-way communication capability. Contract AF 16 439 529 authorized the Pellston Electronics Division of the Pellston Corporation to undertake a development project to reduce the power requirements, the weight, and the overall size of the unit.

Redesign of the transmitter circuit board made possible the achievement of two of the three design goals. The weight reduction requirement necessitated a change from the zinc alkaline battery, used to power the original design, to a mercury battery. This change permits the specified battery life requirement to be met while at the same time reducing the weight by 175 grams. An automatic power control circuit has been added to further improve battery life.

Encapsulation of the circuit within a redesigned case incorporating water sealing gaskets at all seams provides for complete waterproofing of the unit. Overall dimensions of the new case meet or exceed the specification.

Test results demonstrate that all specified performance criteria have been met.

Sample Table of Contents

TABLE OF CONTENTS

-3-

Sample List of Illustrations and List of Tables Combined on One Page

LIST OF ILLUSTRATIONS

LIST OF TABLES

Sample of the First Page of the Report Proper

I. INTRODUCTION

The AN/ART-33 Radio Set has been successfully employed by the Canadian Armed Forces since 1986. As a result of an unsolicited proposal submitted by Pellston Electronics, Aviation Equipment, Ltd., has awarded Contract AF 16 439 529 to Pellston Electronics to further refine the AN/ART-33 design. This report provides all pertinent details of this development.

The project focused on a redesign of the transmitter. This new design provides for an overall reduction in the size of the AN/ART-33, while at the same time requiring less power to achieve the specified performance values.

A new case design and battery type results in a significant reduction in the total weight of the unit.

The report concludes with a discussion of the test procedures and results.

II. DESIGN REQUIREMENTS

The redesign of the AN/ART-33 was governed by Specification CAN-R-9898G, issued by Aviation Equipment, Ltd., on May 17, 1997. Full compliance with all performance specifications has been achieved.

A computer analysis of VHF wave propagation showed that the minimum power output requirements could be met without difficulty. However, thermal compensation of the oscillator bias network was necessary to ensure that the frequency stability specification would be met. Test results show that this method of limiting any tendency to instability in the primary oscillator circuit is completely effective.

No problem of any kind was encountered in meeting the size constraints. A mercury battery was selected to replace the zinc alkaline battery used in the old design. The resultant weight savings brings the overall weight of the unit to 0.975 kg, 25 g less than the specified 1 kg upper weight limit.

III. TRANSMITTER DESIGN

The specified size and weight limitations imposed by Specification CAN-P-7564-TS dictated a rigorous approach to circuit efficiency. Several computer-generated designs were analyzed for power efficiency characteristics. Figure 1 is a block diagram of the most efficient of these designs.

-5-

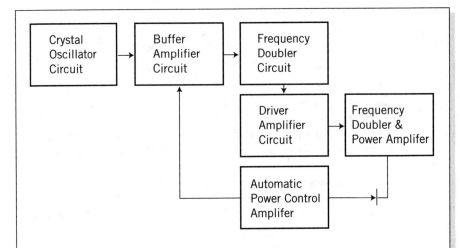

Figure 1 — Block Diagram

A. <u>Oscillator Circuit</u>

The oscillator circuit design is critical in the achievement of frequency stability, particularly within the temperature ranges in which the equipment is required to operate. Therefore, a thermal biasing network was incorporated in the oscillator circuit.

As shown in figure 2, ...

-6-

Exercises

Note: No special technical knowledge is required to complete the exercises in this chapter.

10-1. Writing a Long Report

This assignment should be done by a technical writing group, assembled in accordance with the guidelines given in Chapter 3.

1. Carefully study the following case.

2. In a group meeting, plan the report and prepare an outline for a long report responding to the problem presented in the case.

Note: At this point, the Group Leader should assign jobs based on a division of the outline and the other tasks remaining.

3. After reviewing the chapter and the sample report (pp. 185–192), write a first draft, incorporating a heading structure keyed to the outline and including all the sketches that will become illustrations in the final copy.

Note: One illustration is given in the case. However, at least four more illustrations based on this one illustration are possible and would most certainly be contained in a report of this nature.

4. Rewrite the draft, following the steps given in Chapter 6.

5. Write an abstract in accordance with the instructions given in the chapter.

6. After the page numbers are finalized, prepare a table of contents, list of illustrations, and list of tables. Prepare a cover and title page, and assemble the final document. Finally, write a letter of transmittal to accompany the package of documents to be sent to the customer.

THE LAND RECLASSIFICATION STUDY

Mr. Dobbs telephones you to say that another little technical writing job needs to be done. The Management Consulting Division has just completed a study for a ratepayers group in Washburn County, and the technical writer, swamped with work on another project, needs help. Dobbs wants you to turn the collected information into a long report.

BACKGROUND INFORMATION

A county bylaw proposing to reclassify a large segment of land from "agricultural" to "country residential" has been met with stiff opposition from county residents, farmers and non-farmers alike. In order to make certain that their concerns are properly dealt with, these county residents have formed an association, the Washburn County Residents' Association, which has hired Pellston Management Consulting to carry out a study of the current government services and of the effects the new bylaw will have on them.

The association's four areas of concern are schools, roads, agriculture, and water. Pellston consultant Mario Mattone looked at each of these in turn and made the following notes.

MARIO MATTONE'S NOTES

Two water problems exist, one a supply problem and the other a drainage problem. First, water pressure is not adequate to serve the current population, and if the county permits new residential areas, the problem could get a lot worse. Approximately half of the farmers and two-thirds of the non-farmers living in the area get their water from the county's water reservoir. Half of the farmers and a third of the non-farmers have their own wells. The totals are 4 farms with access to private well water and another 3 that rely on the county reservoir. In addition, 319 of the 488 private homes in the area rely on the county for their water supply. Belmont Farm, the largest in the county, relies completely on the county water supply, as does the 5-BAR-B cattle operation and the corporate-owned sheep farm.

For those residents who do not have private wells, serious water shortages have occurred three times in the last 4 years. Moreover, low water pressure (low enough to cut off the water supply completely to some subscribers) has been common. More than a thousand complaints have been logged in the past 3 years. The residents fear that the new bylaw will permit developers, who have plans for another 530 houses in the county, to build before the problem of an already inadequate water supply has been solved. The 5-BAR-B Ranch had to truck water in to feed stock six times in the past 2 years. The pressure problem does not have to do with low reservoir levels, but rather with an antiquated underground piping system. I talked to Redbone Trenching, and they said that many sections of the system need replacing. They did a cost analysis for the county a year ago. Their estimate for upgrading the system to minimum standards was just under $60 million. County officials decided against going ahead with the program at that time.

The second water problem has to do with drainage. The water table appears to have risen in the past 2 years. The county supervisor said this rise was just a temporary problem caused by several years of heavy rainfall. To be certain, however, I engaged the services of Dixie Star Geophysical,

Inc. They studied the region thoroughly, drilling 15 water table wells and installing 10 piezometers.

They measured water levels in the water table wells and in the piezometers. They also tested for hydraulic conductivity with all piezometers. They tested the rate of discharge from weeping tiles around three houses (widely spaced) in each of the nine subdivisions. A summary of their findings follows:

1. Unless there is a marked, permanent drop in regional rainfall amounts, the problem of a too high water table in this area will be permanent unless something is done to reduce it.

2. The water table levels were lower 4 to 6 years ago, but that was a temporary situation that came on the heels of a 5-year drought.

3. The area on which one large subdivision is built has variable glacial lake silty clay deposits just below the surface. These deposits are from 10 to 16 feet thick all through the subdivision area and they lie over top of a much larger and deeper layer of water-bearing sandy silt.

4. Given the water table levels, seepage from septic fields contains or soon will contain inadequately treated sewage. All the residents of the county have septic tanks, there being no sewer system.

5. There is no storm sewer system either. Excess water (from melting snow or heavy rain) is carried away by drainage ditches, many of which are too shallow to accommodate the runoff from a major rainstorm (more than 2 inches of rain in a 24-hour period) or a sudden spring thaw. As a result, water can overflow and add to the water table level and to the erosion of the roads in the area.

6. The average discharge from the weeping tiles around the houses tested was 0.016 gal/second in Subdivision A, 0.011 in Subdivision B, 0.012 in Subdivision C, 0.011 in Subdivision D, 0.011 in Subdivision E, 0.012 in Subdivision F, 0.009 in Subdivision G, 0.015 in Subdivision H, and 0.014 in Subdivision I.

A layout of the county is shown at the top of page 196.

Several county roads need resurfacing, particularly the one running east and west just to the north of Subdivisions G and C and the one just to the north of the site of the larger of the two proposed new subdivisions. If the new subdivisions go in, these roads will have to bear large increases in traffic.

In addition, the Trans-Canada is a limited-access highway where it passes through the county. There is but one cloverleaf access in the northwest part of the county, and it is impossible to get onto the highway without first getting onto the Powhatan Parkway. Since the city lies to the southeast, where most of the non-farmers work, a second cloverleaf would be highly

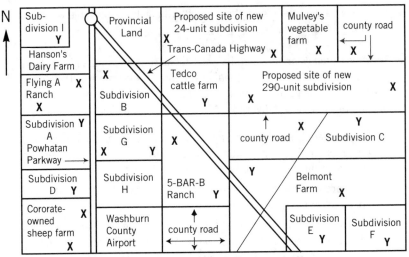

Location of water table wells shown by **X**
Location of piezometers shown by **Y**

desirable. The best place to put this cloverleaf would be at the intersection of the Trans-Canada and the county road that runs north and south between the 5-BAR-B Ranch and Belmont Farm.

This cloverleaf would cost the county $2.1 million, but the wear and tear on county roads would be dramatically reduced. With reduced traffic flow inside the county, the amount of money needed for keeping county roads in good repair could be cut, eventually paying for the cloverleaf. The current budget for road repair is $330,000, a figure about half that needed to keep the roads in good repair. The resurfacing of the two county roads mentioned earlier would cost $4.3 million in addition to the standard repair budget.

There is one school in the county, the William Lyons Elementary and High School. It is located in Subdivision A just off the Powhatan Parkway. About 90 percent of the 600 school-age children in the county attend this school. The rest attend private schools in the city or are enrolled in home-school programs. If the development plans go ahead, at least one more school will be needed. Added property tax revenue from the new residents in the county will add approximately $1.3 million to the $1.75 million currently collected.

I asked the provincial Department of Agriculture to test the soil in the areas of the two proposed subdivision sites. Most of the soil in these areas is Number 1 soil. This type of soil can support crops in all except the worst droughts. Using these sites to put residential housing on doesn't make much sense from the standpoint of soil conservation or proper land use. I think we ought to give some emphasis to this point in our report, since good farm land is such a valuable natural resource. (The current subdivisions are built on lands that are not considered prime farm land.)

10-2. *Writing a Long Report*

This case provides information for a report to be sent outside a company. It is similar in some respects to a contractual report, and therefore, you are encouraged to use the specification given in Chapter 7 as a guide to front-matter elements and page layout. Ideally, this assignment should be done by a technical writing group, assembled in accordance with the guidelines given in Chapter 3.

1. Carefully study the following case.

2. In a group meeting, plan the report and prepare an outline for a long report responding to the problem presented in the case. The following outline will work nicely for this assignment, but feel free to modify it if you can think of a better way to organize the report.

I.   INTRODUCTION
    A. Purpose of Report
    B. Scope of Report
    C. Plan of Development

II. IN-PLANT INVESTIGATION
    A. Engineering Document Review
    B. Performance Review
    C. Hypothesis

III. FIELD INVESTIGATION
    A. Troubleshooter's Report
    B. Installation Check
    C. Hypothesis

IV. FLIGHT TEST
    A. Methodology
    B. Test Results

V.   CONCLUSIONS AND RECOMMENDATIONS

Note: At this point, the Group Leader should assign jobs based on a division of the outline and the other tasks remaining. As a first step, someone should be assigned the job of preparing a table based on the information related to the airline companies, types of aircraft, installation dates, and performance data (some of this same data was given in Exercise 8-2, Chapter 8, and the answer to that exercise will be useful here). This table can be used to support the first hypothesis.

3. After reviewing the chapter and the sample report, write a first draft, incorporating a heading structure keyed to the outline. *Do not fall into the narrative trap.* The case provides information, not material to be quoted. *Use names only in the letter of transmittal, in which you will acknowledge the help you received from Sampson's employees.* Plan the illustrations at this stage. Make sure that each illustration has a figure number and a suitable title. Use all the illustrations from the case, keeping in mind that all block diagrams, line drawings, etc., are illustrations. In other words, you will end up with one sequence of figure numbers, not one for block diagrams and another for line drawings.

Note: Use all the illustrations given in the case. Be certain to refer (within the text) to each illustration by figure number and each table by table number. The illustration or table thus referred to should then follow its reference as closely as possible. Once this requirement has been satisfied, the illustration or table may be referred to again as necessary. Provide appropriately placed captions that include a figure number and title for each illustration and a table number and title for each table.

4. Rewrite the draft, following the steps given in Chapter 6.

5. Write an abstract in accordance with the instructions given in this chapter.

6. After the page numbers are finalized, prepare a table of contents, list of illustrations, and list of tables. Prepare a cover and title page, and assemble the final document. Finally, write a letter of transmittal to accompany the package of documents to be sent to the customer.

7. Write final drafts of the abstract and introduction. After final typing, double-check to ascertain that the correct page numbers are given in the table of contents and in the lists of illustrations and tables. Make certain that no figure number, table number, or page number is larger than any of those that succeed it in the list. (If such a situation does exist, you have made an error in numbering the illustrations or tables.)

8. Do a final reading to pick up any errors you may have missed. Make certain that page changes necessary to correct any errors found during the final reading are accounted for in the table of contents and in the lists of illustrations and tables.

THE AUTOPILOT COUPLER REPORT

You are temporarily assigned to the Engineering Department as a Staff Assistant. However, you are being groomed to work in a field office, and as part of your preparation, you are to familiarize yourself with all of the product lines.

One day, Darla Petoskey, the Chief Engineer, calls you into her office. "Have a look at this," she says, as she hands you a letter.

You note that the letter, dated 27 December 1998, is from Mr. Kermit T. Icarus, President of Sampson Airlines. You glance through the letter quickly, making special note of the paragraph that reads as follows:

The Model 415 Autopilot Couplers, manufactured by Pellston Electronics, Inc., failed in flight in all 10 of the aircraft in which they were installed. The units were all installed in December and, in all cases, failed within the first 2 hours of flight time. Eight of the units failed in the first hour of flight time. The other two failures occurred during the second hour of flight. We request that you immediately address this problem.

When you look up, having finished the letter, Darla continues, "Sampson Airlines is located in the area that you will be serving as a field engineer. You might just as well look into this. It will give you a chance to meet some of the people you will be working with in that region, and it should be a pretty good test of your ability to solve a sticky problem. Go to it."

You decide that your first step should be to check the files for the address of Sampson Airlines, and you discover it is as follows:

Sampson Airlines
1297 Mason Avenue
Whitehorse YT Y81 3M9

Next, you decide to talk to the project engineer on the Model 415, a worried-looking little man named William Broderick. William has already heard about the problem with the Model 415 and seems relieved that someone else has been brought in to try to find a solution. William drags out what appear to be several hundred engineering drawings, manufacturing process sheets, engineering change orders, test procedures, and calibration reports. As he shuffles through these mounds of documents, he talks constantly, pointing to one document and then to another. You listen attentively, trying to follow his line of reasoning, but soon you are lost in the maze. "Hold on, William," you finally say. "Do you know why these things failed?"

"I don't have a clue," he replies desperately. "Every coupler unit is calibrated and tested thoroughly before it leaves the plant, and there is absolutely no evidence of design weakness. Furthermore, these units have operated successfully in 27 other airplanes belonging to two other airline companies."

Finally, you think to yourself, William has given you a valuable piece of information. You thank him and go immediately to the Contracts Department, where the contracts administrator, Hannah Rutledge, helps you to find the appropriate files. You note the following information:

Southern Alberta Airways has 21 Model 415 units. Fifteen are installed in their DeHavilland Dash-7s; 4 units are installed in their Shorts 330s; the remaining 2 units are installed in their Beechcraft 99s. Both of the Beechcraft 99 installations were done in June, but only 5 of the Dash-7 installations were in June. The Shorts 330s weren't fitted with Model 415s until August (1 unit) and September (the final 3 units). The remaining 10 units were installed in the Dash-7s as follows: July (3 units), August (2 units), and September (5 units).

Northern Manitoba Airlines has 6 Model 415s in operation, installed in their Beechcraft 99 fleet as follows: June (2 units), July (1 unit), and August (3 units).

Hannah confirms what William had said: no failures have been reported by either Southern Alberta Airways or Northern Manitoba Airlines. Hannah also tells you that each airplane flies roughly 350 hours per month. A quick mental calculation tells you that these airplanes have had a lot of trouble-free hours from their Model 415 coupler units. Only Sampson has had difficulty with the coupler, and all of Sampson's airplanes are Convair 240s.

Recalling that all of the Sampson installations were done in December, you wonder if a bad batch of units somehow left the plant. Since the documentation search by William Broderick failed to turn up anything, however, you decide that this possibility is too remote for further consideration.

Your next step is to find out as much as possible about the installation procedures. Still in the Contracts Department, you discover that these procedures were subcontracted to Aero Technical Publications, Inc, located in Sunnyside. You make a telephone call to Aero to set up a meeting.

At Aero, you meet Josephine Merkley, the manager, and David Burton, the writer who did the actual work of writing the procedure. Mr. Burton explains that the installation is a little different for each type of aircraft, but that the wiring diagrams of the aircraft are carefully checked to assure that the installation procedures are correct.

Another possibility occurs to you, that of an incompatibility with the autopilot with which the coupler must operate. However, Mr. Burton tells you that engineers at both Sampson Airlines and at Pellston Electronics double-checked everything before approving the procedures.

You learn also that airline technicians working for the individual airlines installed the coupler units. You wonder if a particular installer at Sampson Airlines could have misinterpreted the procedures and connected something incorrectly. You make a note to interview the installer when you travel to Whitehorse to further investigate the failures.

At this point, you decide that you have about all of the information you are going to get without going to Sampson Airlines.

You tell Darla Petoskey that you are going to Whitehorse to try to solve the problem. "Good idea," she says. "Make arrangements to get up there right away. Get to the bottom of this thing. I'll phone Icarus and tell him you're coming. We'll have to send a full-scale report of our findings to them, so keep that in mind."

After an uneventful trip the next morning, you check into your hotel in Whitehorse. You head for Sampson's maintenance hangar. Darla's phone call to Mr. Icarus has prepared the way, and you are greeted cordially by Terry Knot, Chief Maintenance Engineer. Being a no-nonsense type of person, he ushers you immediately to the avionics bench area, where he introduces you to an avionics technician whose name is Moss Phegt. "Moss," Terry Knot says, "will you tell the Pellston representative here what you have found out about the coupler failures?" He then says good-bye, leaving the two of you to discuss the problem.

Moss shows you a unit that is sitting on his workbench. "I dug this one out of the storeroom so that you could look at it. They're all the same anyway."

"Do you mean that all of them failed in exactly the same way?" you ask.

"Yes," he replies. "Do you know what this thing does?"

"Yes," you say, "but maybe you'd better go over it anyway," not wanting to miss any bets.

"Well," he says, "the coupling unit — this thing right here that's made by your company — has seven circuit boards. The circuitry on these boards drives the pitch controller, converts information from the navigational equipment, that is, the VOR, ADF, and GPS, to control the roll and yaw for correct headings."

"That's what I thought," you say. "These are the circuits that actually provide the information used by the autopilot system?"

"That's right. These circuits fly the airplane in that sense. Most of these circuits feed information to the autopilot to keep the airplane on course. They pick up signals from the outputs of the navigational equipment and process it so that the autopilot can direct the airplane to a ground station. There isn't any problem with any of them. [See Figure 1 at the top of page 202.]

"Now, if you will look right here at Board 7," he says, pulling the board out of the unit, "you'll see that this integrated circuit on the edge of the

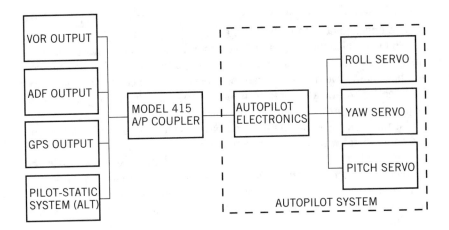

Figure 1 — Block Diagram of Coupler Function

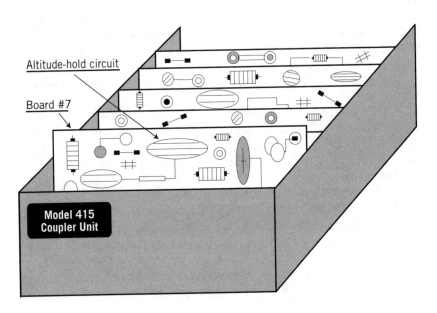

Figure 2 — Pellston Model 415 Coupler Unit

board somehow failed. [See Figure 2 at the bottom of page 202.] The function of that particular circuit is to provide information for operation of the pitch controller in the autopilot. It keeps the airplane at its assigned altitude so that slight downdrafts or updrafts will not cause the ship to go up and down in the air.

"When this here altitude-hold circuit fails, false information gets into the pitch servo of the autopilot system, and that will cause the airplane's nose to pitch up violently. The pilots are able to regain control, but they have to act fast. The first thing they have to do is to switch off the coupler unit. After that they can re-engage the autopilot. Of course, without the coupling unit operating, they have to keep the airplane on course by hand and watch for any altitude changes."

Now, you have a fairly complete picture of what happened. When the coupler processed the altitude information incorrectly because of the failed circuit, the coupler had to be switched off. The pilot then had to fly the airplane by visual reference to the information shown on the navigational instruments and the altimeter. But you still don't know why that circuit failed. Looking in your notebook, you see the note to remind you to interview the installer.

The installer turns out to be a middle-aged woman named Gloria Grunwald. Carrying a copy of the installation procedures provided by Ms. Merkley of Aero, you and Gloria climb aboard a Convair 240 that is in the hangar for a maintenance check. You and she go step-by step through the procedure, and you learn that she has adhered strictly to the procedures provided.

You seem to have come to an impasse: the coupling unit itself is apparently not at fault; engineers have ruled out the possibility that the autopilot system and the coupler are incompatible; the procedures have been checked and rechecked; and the installer has done a proper job of installation. Your only hope now seems to be to see if the troubleshooter, Moss Phegt, can offer a theory.

You go back to the avionics maintenance bench to talk to Moss once again. "You say this circuit failed, Moss. What do you think caused it to fail?"

"Well," Moss says, "that circuit is on the edge of the board. If it came in contact with the top cover, and if the top cover was hot, I suppose the circuit could have overheated. That's the only thing I can think of. Still, that shouldn't have caused the problem — unless the top cover got pretty hot. Let's take another look at the installation."

You follow Moss Phegt into the Convair sitting in the hangar. Even though you and Gloria just removed and re-installed the unit a few minutes ago, you try to see something you might have missed before. "Here's the layout," Moss says, indicating the Forward Electronics Rack. [See Figure 3 at the top of page 204.]

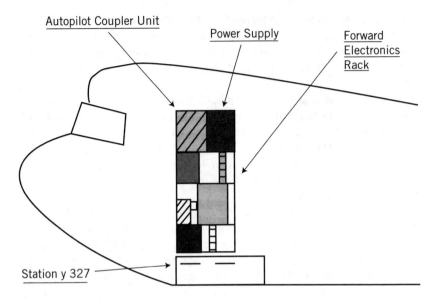

Autopilot Coupler Unit

Power Supply

Forward Electronics Rack

Station y 327

Figure 3 — Installation of Model 415 Coupler in Forward Electronics Rack

You look at the arrangement of units in the console and notice that the power supply is right behind the coupler unit. "I'll bet this power supply generates a lot of heat," you say.

Moss frowns slightly. He squints, suspiciously now, at the close proximity of the power supply and the coupler unit. "I suppose that a fair amount of heat could be transferred through here," he says, trying to poke a screwdriver between the power supply and the coupler unit. "In fact, this top cover could be acting almost as a heat sink for the power supply."

Finally, you have come up with a possible cause of the failures. "Moss," you ask, "could you figure a way to insulate these two units?"

"Sure," he says, "I can put a new circuit in that unit on the bench, throw a layer of insulation between the boards and the top cover. Then you can take her up in the air and see what happens," he adds with an evil grin.

The next day, you find yourself 7,000 feet up with an unobstructed view of the stars. You are seated next to Harry Derring, Sampson's check pilot, who has been flying you back and forth across the tundra for three hours, the coupling unit activated. "Okay, Harry, switch it off," you say.

After Harry turns off the coupling unit, you disconnect it, take it out of the equipment rack, remove the top cover, and remove the layer of insulation that Moss Phegt has placed there. Then, you replace the cover without the insulation and re-install the unit in the equipment rack. You come back into the cockpit and settle into the right-hand seat. "Okay, Harry, switch it on."

Harry reactivates the coupler, and you sit back and wait.

Twenty minutes later, you get a funny feeling in your stomach, as the stars seem to roll downward in front of you. You realize that the Model 415 Coupling Unit has failed and that the nose of the Convair has pitched up. Harry quickly regains control of the airplane, and you will live to write the report, having been provided with strong evidence that your theory is correct.

CHAPTER 11

External Proposals

OBJECTIVES: In this chapter, you will discover the differences between solicited and unsolicited proposals. You will also learn how to manage a proposal project, including all of the steps, from planning to final production.

What Exactly Is a Proposal?

Although it shares some of the surface characteristics of a report, a proposal is not a report. Its foundation and purpose are completely different from those of the report. Proposals written in response to a Request for Proposal (RFP) are called *solicited proposals*. Those written to provide a product or service identified by a vendor, rather than by a customer, are called *unsolicited proposals*. In both cases, the purpose of the proposal is the same: to offer a product or service to a potential customer.

Writing a report on a completed project or on a development already under way is usually not easy, but writing a proposal on a development or a study that has yet to begin presents additional difficulties. For example, if the proposal is solicited, some engineering and preliminary study will probably have to be done as part of the proposal project. Unsolicited proposals also sometimes require engineering effort, even if the proposal has grown out of a completed program. And although a report writer may be able to pull together all of the technical data necessary to write a report, a solitary proposal writer who hopes to produce a proposal on time (deadlines are nearly always imposed on proposal submittal) may face a nearly impossible job. Fortunately, proposals of any consequence are always produced by a team of people, each of whom has special qualifications and responsibilities.

Some RFPs include a request that the technical proposal not mention cost. Customers include this provision so that their technical personnel will be able to evaluate each proposal solely on its technical merit. In such cases, a cost and contractual proposal, bound separately, is prepared for evaluation by the customer's financial experts. Some customers may also request a third volume called a management proposal to be evaluated by their administrators.

Short proposals, on the other hand, can be done in a letter format (see Chapter 9). Proposals following a letter format, like their report counterparts, are written in an objective, non-narrative style. To this end, proposal writers use personal pronouns sparingly or not at all.

A Typical Proposal

A typical full-scale proposal contains the following parts:

Volume 1: Technical Proposal

- Cover, die-cut or preprinted
- Letter of transmittal
- Title page
- Abstract or summary
- Table of contents
- List of illustrations
- List of tables
- Technical sections, including work statement, etc.
- Boilerplate sections
- Appendices, as required

Note: The term *boilerplate* is used to refer to those proposal sections that discuss matters such as company facilities, experience on similar projects, personnel qualifications, and anything else that does not specifically refer to strictly technical matters.

Volume 2: Cost and Contractual Proposal

- Cover, die-cut or preprinted
- Title page
- Foreword or abstract
- Contractual information
- A certificate of bidder representation
- Certified cost data breakdown

Elements of a Technical Proposal

THE PROPOSAL COVER

The same type of die-cut cover used for technical reports is appropriate for use as a proposal cover. Sometimes, companies will print special covers for proposals, in order to make their proposals especially attractive. However, most companies use standard covers that provide company identification at a glance.

THE LETTER OF TRANSMITTAL

A letter of transmittal accompanies a package of proposal copies for the same reason that one accompanies a package of report copies (see Chapter 10). The proposal's letter of transmittal, particularly if the proposal is solicited, may differ markedly from that accompanying a report. The proposal's inherent sales characteristic may dictate that a positive statement of the proposal's responsiveness to the RFP be included, even though this statement will be made again in somewhat different terms in the abstract and in the statement of work.

Also, acknowledgments are frequently unnecessary in a proposal's letter of transmittal. The final paragraph usually provides assurances that the project will be executed within the specified time and budget constraints. (Please refer to the letter of transmittal in the sample proposal appearing at the end of this chapter, p. 217.)

TITLE PAGE

A proposal's title page is similar to that of a report. It may be configured in such a way that the title shows through the die-cut window of a standard cover. If the proposal is solicited, the number or other designation of the RFP must be included as part of the title. If multiple volumes are needed, all titles should be the same except that *Technical Proposal for, Management Proposal for,* or *Cost Proposal for* will begin their respective titles.

Other information will include the name of the vendor's company and that of the receiving company or agency. Of course, all title pages must also contain a date.

ABSTRACT

The abstract of a proposal is no different from the abstract of a report. It should provide a brief summary of the contents of the proposal.

TABLE OF CONTENTS

The table of contents of a proposal must provide a complete listing of all headings, along with their corresponding page numbers. In tables of contents of reports, it is sometimes desirable to reduce the size of the list by eliminating headings beyond levels 2 or 3. However, this practice should not be carried over to tables of contents written for proposals. Proposal readers must evaluate and compare your proposal with those of other bidders, and eliminating items from the table of contents makes this job harder than it needs to be.

LIST OF ILLUSTRATIONS AND LIST OF TABLES

An accurate list of illustrations also makes the proposal evaluator's job easier than it would be otherwise. Be certain to include all illustrations, along with a triple-checked page number for each. Proposal production must often be hurried, but the few minutes one takes to carefully check the accuracy of all front-matter items are well spent.

TECHNICAL SECTIONS

A proposal's technical sections must be carefully organized and clearly presented. Technical discussions meant to impress the reader with the writer's erudition, rather than to clearly express the concepts and designs that will lead to a successful program, have probably cost more companies more money in lost contracts than has any other single error.

BOILERPLATE SECTIONS

The term *boilerplate* refers to the sections describing facilities, experience on similar projects, personnel qualifications, project organization, and anything else that does not specifically refer to strictly technical matters. Most companies use "canned" write-ups that can be patched together as needed for each individual project. For example, if a company were proposing to develop a piece of airborne equipment that must operate at high altitudes and low temperatures, it would include descriptions of its environmental test laboratory; if it were bidding on a large manufacturing run, it would tell of the space and machinery available; and so on.

Likewise, descriptions of past experience on projects that have relevance to the project proposed would be described in some detail.

The project organization should include a chart showing the people who will occupy key positions in the project should the proposal be successful, along with brief descriptions of the qualifications of each.

As mentioned earlier, however, very large proposals sometimes contain the boilerplate sections in a separate volume called a *management proposal*.

APPENDICES

Items to be included in appendices include long mathematical explanations, test procedures and measurements, and all other peripheral supporting information that would clutter up the text if it were included in the proposal proper.

Contents of Other Volumes

COST AND CONTRACTUAL PROPOSALS

Cost and contractual proposals are often submitted as separate volumes so that the evaluation of the technical proposal will be unaffected by monetary matters. These volumes are usually written by non-engineering people who "crunch" the numbers based on the information supplied by the line departments that will execute the program. Although engineers and technicians have to estimate the number of hours they or their staff will likely spend on certain operations, the cost and contractual proposal is usually written by contracts administrators and accountants.

MANAGEMENT PROPOSALS

Management proposals are required only for proposed programs that are very large and complex. Because the proposal for a complex program is itself often very complex, two volumes, comprising a technical proposal and a management proposal, are sometimes submitted. The boilerplate sections are shifted from the technical proposal to the management proposal in such cases.

Proposal Management

Regardless of a proposal's size, proposal management is of utmost importance. Two other aspects of proposal preparation also require special attention: responsiveness to technical requirements and correct tone. (Please refer to the sample technical proposal that appears at the end of this chapter, p. 215.)

The first principle of good proposal management is that one person, a proposal leader, must have the authority to set schedules and impose deadlines. Ideally, the person designated as proposal leader will know how long it will take to write, edit, approve, print, collate, bind, package, and ship the copies. The greatest ideas in the world are completely useless if the proposal doesn't get to the customer on time. All RFPs specify closing dates, beyond which proposals will not be considered. Government agencies, which must observe fair-trading practices, are particularly demanding in this regard.

Once designated, the proposal leader must have absolute authority to oversee the proposal's production. Many proposals fail because this simple principle is ignored.

Boilerplate sections must be written ahead of time. This work can be made easy if some system of information storage and retrieval is in place. In an ideal situation, this material is electronically stored for easy access. Pertinent information can then be selected and minor changes made so that these parts of the proposal can be dispensed with early in the proposal effort.

A preliminary plan for a proposal might look like the following list:

May 17	Proposal due at customer's facility
May 16	Six copies each of technical proposal and cost and contractual proposal shipped by courier
May 15	Final check
May 13–14	Weekend
May 8	Camera-ready copy to print shop
May 8	Final corrections to camera-ready copy
May 6–7	Weekend (can be worked in an emergency)
May 1–5	Final editing and approvals
April 17	Edited draft to typist for word processing and laser printing of camera-ready copy
April 10–17	Rewriting and editing of technical contributions
April 10	Contributions from all members of proposal team to proposal leader
April 3–10	Writing of first draft of technical sections by team members
April 7	Boilerplate sections to print shop
April 6	Final editing and approval of boilerplate sections
March 6–31	Preparation of camera-ready copy for all boilerplate sections
March 31	Engineering complete
March 1	Final outline approved and all engineering and writing tasks assigned
February 24	Decision to bid
February 15–24	Preliminary study of RFP

Figure 11-1 shows the foregoing planning list converted to a flow diagram.

FIGURE 11-1	Typical Flow Diagram for Proposal Project

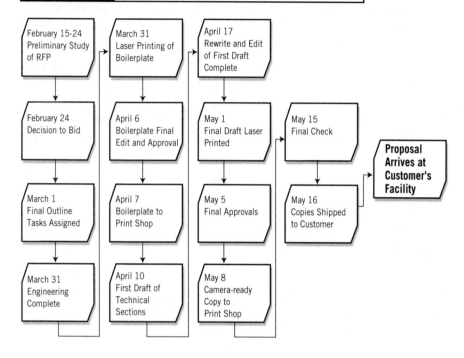

February 15-24
Preliminary Study
of RFP

March 31
Laser Printing of
Boilerplate

April 17
Rewrite and Edit
of First Draft
Complete

February 24
Decision to Bid

April 6
Boilerplate Final
Edit and Approval

May 1
Final Draft Laser
Printed

May 15
Final Check

Proposal
Arrives at
Customer's
Facility

March 1
Final Outline
Tasks Assigned

April 7
Boilerplate to
Print Shop

May 5
Final Approvals

May 16
Copies Shipped
to Customer

March 31
Engineering
Complete

April 10
First Draft of
Technical
Sections

May 8
Camera-ready
Copy to
Print Shop

Responsiveness

Meeting the customer's requirements is the most important job of any proposal. First, the proposal must indicate that the project team clearly understands the work to be done. The simplest method of accomplishing this step is to include a "Statement of Work" that reflects the customer's expectations and that is cast in the same words used in the RFP.

Occasionally, exceptions must be taken to one or more sections of an RFP. Proposals should clearly delineate exceptions and the consequences of each exception. When no exceptions are taken, the proposal should contain a statement to that effect.

If several approaches to the customer's problem are possible, the technical sections should discuss the relative merits of each one, along with a full disclosure of the reasons for the approach chosen. Although the merits of one approach over another may be obvious to the proposal writers, the proposal evaluators may not have analyzed the problem to the same extent. Discussion of alternate approaches anticipates questions such as the following:

Why didn't they follow this other method, which certainly seems simpler and cheaper?

In some cases, an improved version of the product or service requested can be offered for little or no additional cost. Proposals containing this kind of an alternative often enjoy an advantage over the competition.

Customers expect assurance that the proposing company has all the facilities, experience, and support necessary to carry out the proposed program. The proposal should show that the company has all of the necessary qualifications.

Tone

Proposals that contain jargon, poorly structured sentences, and lengthy paragraphs show a disdain for the reader. The tone produced by stilted writing is completely inappropriate in any technical writing, but it is especially so in proposals.

Proposal evaluators are impressed by clear writing, not by simple ideas conveyed in complex sentence structures. Clear writing is an indication of clear thinking, which produces sensibly designed products and services. Writing that exhibits muddy thinking or dishonesty will not likely sell a customer on the proposed ideas.

Although a proposal is a sales document, its power to persuade does not derive from that which we have come to know as sales methods. The proposal should inform rather than promote, analyze rather than argue, and show rather than tell.

In other words, the word *technical* is the operative word in *technical proposal*. No proposal evaluator would take a proposal seriously if it were presented in a sales or promotional writing style.

The following sample proposal illustrates the elements of a technical proposal.

Sample cover with die-cut for title

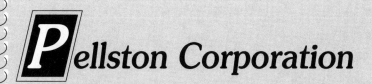

Pellston Corporation

> **Technical Proposal**
> **for**
> **RADIO SET AN/ART-33**
> **DEVELOPMENT PROJECT**
>
> June 10, 1998

Quality Electronics Systems

An Industry Leader
since 1958

Page 1 of 51 pages

Technical Proposal

for

RADIO SET AN/ART-33
DEVELOPMENT PROJECT

June 10, 1998

Submitted to:

AVIATION EQUIPMENT, LTD.
Halifax, Nova Scotia

In Response to

RFP 89-5422XP

Submitted by:

PELLSTON ELECTRONICS
DIVISION OF
THE PELLSTON CORPORATION
Sunnyside, Ontario

ellston Corporation

4567 Airport Road
Sunnyside ON K2V 1M8

Telephone: (613) 555-5000
Fax: (613) 555-5333

An Industry Leader Since 1958

June 10, 1998

Dr. Joan B. Gonzales
Contracts Administrator
Aviation Equipment, Ltd.
5567 Armbruster Avenue
Halifax NS B8H 9Y9

Dear Dr. Gonzales:

Twelve copies of the Technical Proposal for AN/ART-33 Development Project are herewith submitted in response to Aviation Equipment, Ltd., Request for Proposal 89-5422XP.

No exceptions are taken to the performance specifications included as part of the RFP.

We are confident that our proposed program represents the best course of action for transforming the AN/ART-33 into an easily manufactured unit to meet Specifications CAN-R-9898G and CAN-P-7564-TS.

We would be pleased to answer any questions that may arise in connection with our proposal.

Yours truly,

THE PELLSTON CORPORATION
ELECTRONICS DIVISION
Fowler Moffat
Marketing Manager

Enclosure: Twelve copies of "Technical Proposal for AN/ART-33 Development
            Project"

Sample Abstract

ABSTRACT

In Response to Aviation Equipment, Ltd., Request for Proposal 89-5422XP, the Pellston Electronics Division of the Pellston Corporation proposes to undertake a development program to reduce the power requirements, the weight, and the overall size of Radio Set AN/ART-33, a solid-state emergency locator transmitter with a two-way communication capability.

We believe that the redesign of the transmitter circuit board will make possible the achievement of two of the three design goals: the power and size reductions. The weight reduction requirement can be met by changing the currently used zinc alkaline battery to a mercury battery. This change will ensure the specified battery life and at the same time will reduce the weight by more than 100 grams. An automatic power control circuit will be added to extend battery life beyond the specified minimum.

Encapsulation of the circuit within a redesigned case incorporating water-sealing gaskets at all seams will provide for complete waterproofing of the unit. No difficulty in meeting the specified overall dimensions of the new case is anticipated.

-2-

Sample Table of Contents

TABLE OF CONTENTS

continued →

-3-

Sample List of Illustrations and List of Tables Combined on One Page

LIST OF ILLUSTRATIONS

continued ➤

LIST OF TABLES

Table	Title	Page
1	Component Values	11
2	Performance Data	19
3	Subassembly Weight Analysis	26

-4-

Sample Statement of Work

I. <u>STATEMENT OF WORK</u>

  A. <u>Materials Specification Compliance</u>

In accordance with Aviation Equipment, Ltd., Request for Proposal 89-5422XP, the Pellston Electronics Division of the Pellston Corporation proposes to undertake the redesign of Radio Set AN/ART-33. The newly designed unit will meet all the requirements set forth in Specification CAN-R-9898G, which forms a part of the subject RFP, as follows:

<u>CAN-R-9898G Paragraph</u>

1.0 through 2.10	Full compliance
3.0 through 6.17	Full compliance
7.0 through 9.11	Full compliance
9.12	Does not apply, since Pellston's design obviates the need for forged parts.
9.13 through 12.14	Full compliance

  B. <u>MTBF Calculations</u>

Failure rate data for each individual component will be used to calculate the MTBF. Preliminary figures indicate that the MTBF will approach 8,000 hours, 60 percent beyond the 5,000-hour MTBF specified in the RFP.

continued →

C. Performance Specifications

The redesign of the AN/ART-33 will be governed by Specification CAN-P-7564-TS, issued by the Bureau of Naval Weapons on May 17, 1997. Full compliance with all performance specifications is forecast.

A computer analysis of VHF wave propagation shows that the minimum power output requirements can be met without difficulty. However, thermal compensation of the oscillator bias network will be necessary to ensure that the frequency stability specification will be met. Preliminary test results show that this method of limiting any tendency to instability in the primary oscillator circuit is completely effective.

No problem of any kind is anticipated in meeting the size constraints. A mercury battery will replace the zinc alkaline battery used in the old design. The resultant weight savings will bring the overall weight of the unit to 0.975 kg, 25 g less than the specified 1 kg upper weight limit.

-5-

A Typical Page from a Technical Discussion

II. TRANSMITTER DESIGN

The specified size and weight limitations imposed by Specification CAN-R-9898G dictate a rigorous approach to circuit efficiency. Several computer-generated designs will be analyzed for power efficiency characteristics. Figure 1 is a block diagram of the most promising of these designs, according to preliminary study.

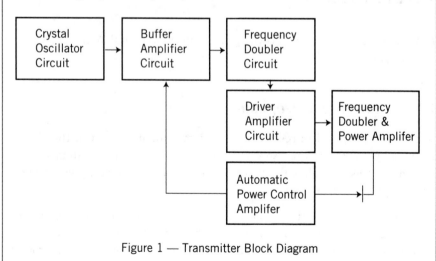

Figure 1 — Transmitter Block Diagram

continued ➤

A. Oscillator Circuit

The oscillator circuit design is critical in the achievement of frequency stability, particularly within the temperature ranges in which the equipment is required to operate. Therefore, a thermal biasing network will be incorporated in the oscillator circuit.

-6-

The Beginning of a Typical Page from a Boilerplate Section

IV. STATEMENT OF CAPABILITIES

A. Facilities

A full range of engineering and test facilities is available for the proposed project. The Pellston Electronics Division occupies 32,515 m² of floor space. Approximately 30 percent of the total space is devoted to engineering and developmental testing laboratories.

Exercises

Note: No special technical knowledge is required to complete the exercises in this chapter.

11-1. Planning a Proposal

Although a proposal in response to a procurement request of the size demanded by the RFP in this exercise would be far more complex than a correct response to this case, all of the elements and problems associated with proposal writing are present.

1. Assemble a technical writing team and begin by reading through the following case study several times, making certain that all members of the team thoroughly understand it. Elect a group leader, and prepare a proposal plan and flow chart, using the model given in the chapter (see p. 212–13).

2. As a group, outline the proposal. Since the RFP asks for a specific organization, most primary headings are already decided. You will want an introduction to precede these headings and possibly a conclusion to follow them. At this point, the group leader should assign the parts to group members.

Note: Although the case study given in Exercise 11-3, which concludes this series of exercises, contains information, it is not yet in first-draft form, and the group leader should assign team members to write first drafts of each section of the outline.

Darla Petoskey has yet another challenge for you. "As you may know," she begins, "Pellston has been a pioneer in geological surveying system development. We have designed and built systems for all sorts of data gathering and processing. In the past few years, we have even expanded into undersea data-gathering system design.

"Environment Canada is now looking for someone to design and build a string of automatic weather stations to be deployed in remote sites in the Yukon and Northwest Territories. We have the staff, and we have the facilities. I want you to head up the proposal effort on this. Here is a copy of the RFP and the minutes of a proposal meeting we had last week. Study them, get in touch with the appropriate people, and get started."

Your first task is to familiarize yourself with the terms the RFP.

ENVIRONMENT CANADA RFP 90-12

Introduction

Environment Canada herewith solicits proposals for the design, development, and construction of Automatic Weather Stations for deployment in remote sites in the North. Following prototype approval, 24 such units will be procured for deployment.

Bidders must submit technical proposals to arrive at Environment Canada, 888 Blankenship Street, Ottawa ON K5C 7X7, no later than 12:00 P.M., March 15, 1998. All technical proposals should be marked for the attention of Ms. Hortense C. Pembroke, Contracts Administrator. No proposal received after the closing date will be considered for this procurement.

Costs should not be included in the technical proposal. A cost proposal, to be solicited at a later date, will be required only of those bidders whose technical proposals are judged to have satisfied the requirements of the subject procurement.

All technical proposals will contain, as a minimum, a work statement, a technical description of the proposed design, a description of the facilities available, and a description of the proposed project organization, including résumés of key personnel.

Technical Requirements

The Automatic Weather Station, hereafter referred to as AWS, will be remotely placed at various sites in the North. The units must be capable of continuous operation for six months without maintenance. Although deployment of the units will be the responsibility of Environment Canada , the successful bidder will be required to furnish complete installation instructions. In addition, the successful bidder will provide support in the form of one field technician for the first year of operation, in order to train Environment Canada personnel in the proper maintenance routines. One prototype unit, accompanied by a full set of test procedures, will be required for submittal 6 months after contract award.

The AWS will be capable of continuously measuring wind direction, wind speed, air temperature, and barometric pressure. The data thus collected will be telemetered to a collection site, at which it will be processed for distribution to staffed weather stations.

Power Requirements

A design that employs solar energy is desirable. However, Environment Canada recognizes that some other method of power generation would have to be used during the winter months. Therefore, consideration will be given to methods of power generation that do not rely on direct conversion of solar energy.

Environmental impact studies indicate that an accident involving any form of nuclear-fueled power-generating equipment, including isotope-filled thermoelectric generators, could damage the fragile environment of the Arctic tundra. Therefore, no power-generating scheme that includes such methods will be considered.

The Platform

The AWS will be placed on the Arctic tundra. Some method for securing the unit to the site must be devised. Also, some method for leveling the device will be required. The materials used should be light enough to obviate the need for transport on a vehicle larger than a Nodwell-type track machine.

The Instruments and Superstructure

Data-sensing devices capable of withstanding Arctic cold and gale-force winds will be required. Superstructure design may be able to compensate for some of the harshness of the environment. In any case, every effort must be made to ensure that the instruments are reliable and capable of continuous operation in an Arctic environment.

The air temperature sensing device shall be capable of sensing temperatures in the range of –62 degrees C to +40 degrees C to an accuracy of 0.1 degree across the entire range.

The barometric pressure measuring device shall be capable of measuring pressure to an accuracy of plus or minus one one-hundredth inch of mercury within the range of 27 to 31 inches of mercury.

Wind direction shall be measured by a sensor capable of giving accurate and continuous readings.

Wind speed measurement sensing schemes shall take into account the gale-force conditions that can obtain in the north.

The rest of the RFP gets pretty technical, and so you skip ahead to the minutes of the proposal meeting.

Minutes of Meeting Held Thursday, November 8, 1997

Present:

Brian Patrick, Advanced Projects Manager

Rita Waechter, Project Engineer, Research and Development

Jean Wilberforce, Technologist, Research and Development

Nicholas Anthony, Engineer, Mechanical Section

Daniel Edwards, Engineer, Communications and Telemetry Section

Colin Andrews, Engineer, Communications and Telemetry Section

Brian Patrick: Ladies and Gentlemen, I think this RFP is just what we have been looking for. I think we have all of the expertise necessary to successfully carry out this project. What do you think, Rita?

Rita Waechter: I can't see any problem, except possibly the telemetry gear.

Daniel Edwards: We're going to have to come up with a pretty rugged unit all right.

Colin Andrews: We can put the sensitive electronics down inside the superstructure, perhaps somewhere near the power generator, for temperature stability. The big problem, as I see it, will be to keep the whole unit from icing up or blowing away.

Nicholas Anthony: I think we can come up with a structure that will stand up to anything Mother Nature can dish out. We'll have to work together closely with Colin. Who is going to design the power unit, Rita?

Rita Waechter: I'm going to be responsible for that myself, but Jean will be helping and doing a lot of the research to help come up with some kind of suitable system.

Brian Patrick: All right, it looks as though we've got our team, and it looks as though we've got our goals. I'll get Garth to appoint a proposal leader right away, and we'll get going on this thing.

Your first responsibility seems to be to get together with Rita Waechter, the Project Engineer, to plot a strategy. In an effort to coordinate the engineering and the writing, you first draw up a preliminary plan for the proposal. (Draw up a plan similar to the one given in the chapter, and then convert it to a flow diagram as shown in the chapter.) After consultation with Rita, you come up with the following block diagram for the project itself.

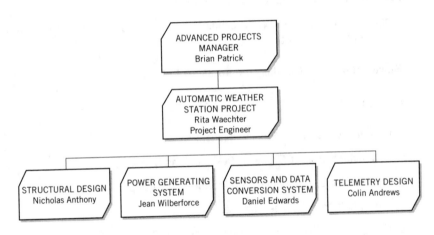

11-2. Writing the Boilerplate Sections for a Proposal

Although a proposal in response to a procurement request of this size in the real world would be far more complex than a correct response to this case, all of the elements and problems associated with proposal writing are present.

1. Carefully study the information in the following box, and outline it in a logical fashion. Write first drafts for the boilerplate sections of the proposal you planned in Exercise 11-1 in accordance with your outline. Plan the illustrations for these sections at this stage. Make sure that each illustration has a figure number and a suitable title.

Note: Use the block diagrams of the company and division organizations (given in the introductory material for Section III, The Forms), along with the project organization given in Exercise 11-1. Be sure to place figure numbers in the text just before the spaces in which illustrations will be placed. Indicate a figure number and a suitable caption for each illustration. Pictures of the environmental equipment are not included, but you may include sketches of your own as an option.

2. Write at least a paragraph to introduce each of the categories of equipment in the environmental test laboratory and then list each piece for clarity and quick reference.

3. When writing the final draft, rearrange the material as necessary for a coherent picture of the company's capabilities.

Note: Keep in mind that all block diagrams, line drawings, etc., are illustrations. In other words, the final proposal will contain one sequence of figure numbers, not one for block diagrams and another for line drawings.

Having been assigned the boilerplate sections, you go to the company library to see what information you can find on the company in general. From some old annual reports, you discover that the company was founded in 1929 by Waldo T. Claypool. The company started out as a supplier of speedometers and fuel gauges to the automobile industry, got into electronics products in the mid-1930s with a burglar alarm design, and gradually worked its way into a share of the vacuum tube market.

From the most recent annual report, you discover also that the corporation has just the three major divisions you already knew about: electronics, musical instruments, and consulting. The common stock, listed on the Toronto Stock Exchange, is held by approximately 10,000 shareholders. Total company assets exceed $900 million, and the company credit rating is AAA-1.

Note: For more information about Pellston Corporation, see also the introductory material for Section III, The Forms.

Another section of the annual report discusses the engineering and engineering support facilities in a general way. The replacement cost of the laboratory equipment within the Electronics Division alone is more than $6 million. It includes a complete range of general and highly specialized instruments with calibration capabilities for the equipment itself. In addition, a prototype construction shop equipped with a full complement of machine tools is available for constructing prototypes and preproduction models. Full computer access is present on the premises for every purpose, including computer graphic representation of developmental designs. In fact, all the engineers have personal computers with CAD (computer assisted design) software suited to their specific specialties. All of the computers are interconnected into something called a LAN. You discover that LAN stands for local area network, which makes communication between the engineers very easy.

After reading more about the CAD software, you find that a design can be completely analyzed by computer, making for great savings in time and materials. Before Pellston acquired this capability, several versions of a design often had to be built and tested before a suitable one was found. Now, only one prototype version has to be constructed because preliminary tests of every kind have been run on the computer. You make a final note that the combined investment in just these computers and accompanying software amounts to $25 million.

Other support facilities include a circuit board laboratory, an extensive library, and an environmental test laboratory. Since the proposal has to do with a weather station, you know that this laboratory is going to be of special interest and decide to investigate further by having a first-hand look.

The laboratory appears as a maze of tanks, chambers, and tables; you therefore introduce yourself to a man in a white lab coat.

"I'm Greg Geiger," he says cheerfully. "How can I help you?"

You explain your problem.

"Generally," Greg begins, "we can test in specified environments of altitude, temperature, altitude and temperature together, temperature and humidity together, shock, linear acceleration, vibration, salt water, and sand and dust."

"Test in those environments?" you ask. "What do you mean by that?"

"Well," Greg continues, "it's not enough to find out whether our products can merely endure a given environment and work satisfactorily afterward. Many times, the equipment will have to actually operate in those environments."

"I see. A radio we produce might have to actually operate in a vibrating airplane at 50,000 feet."

"Exactly," he says. "Let me show you some of the machines. You can make notes as we go along. Here we have a Lorraine Model 43 Altitude/Temperature Chamber. Its limits are –73 degrees C to +93 degrees C and sea level to 70,000 feet. Next, we have a simple bell jar just like the one you probably had in your physics lab at school. It is 12 inches by 8 inches and can be evacuated to simulate the atmosphere at levels up to 70,000 feet.

"Next, we have a Nagel Humidity Chamber with limits of –20 degrees C to +80 degrees C and relative humidities from 60% to 98%. Our shock machine is a Bumper 120 with a 30-foot rail system, and next to it here is a Whipper Model 460 Linear Acceleration Machine with a force of 15,000 g pounds.

"We have three vibration tables: The mechanical one is a Fanshawe, Model B. It has a range of 10–55 cps with a 1/8-inch maximum double amplitude. The two vibration tables with electronic power supplies are Voltco Type A Exciters, both with one-inch double amplitude and a rated force of 1,000 pounds. They'll both go up to 2,000 cps, as well."

You then ask him about the sand and dust and salt water spray equipment.

"Oh yes," he says, "those two facilities are in a room by themselves across the hall. They were both designed and constructed right here at Pellston."

"Anything else?" you ask.

"Well, as you can see, we have a complete range of electronics test equipment that can be rigged up to any device being tested. And the Quality Assurance Department has a similar setup to ours.

Back at the company library, you discover in an appendix to the latest annual report that the total cost in equipment for testing purposes is over $12 million.

After leaving the library this time, you go to the personnel office and gather the following information from the personnel files:

Rita Waechter has a bachelor's degree in electrical engineering from the University of Toronto (1979) and a Master of Business Administration degree from Rutgers University (1985). She has been a Project Engineer with Pellston since 1985. From 1979 to 1984, she worked for Canadair, designing electrical systems for aircraft, specializing in electrically operated hydraulic control systems.

Nicholas Anthony has a bachelor's degree in mechanical engineering from the University of Manitoba (1981) and a master's degree from McGill (1983). He joined Pellston in 1983, and he has worked on a variety of projects, including the mechanical design for an oil field data-gathering system housed in a trailer, which he also designed especially for the purpose. In addition, Mr. Anthony designed a floating platform for an oceanographic survey system. In both cases, highly sensitive instruments were housed in the structures, which had to be designed to accommodate and protect them.

Jean Wilberforce received an honors diploma in electronics from the Northern Alberta Institute of Technology in 1982, at which time she joined the Pellston Corporation. She has specialized in remote power generating devices for a variety of purposes, including the oil field and oceanographic projects that Nicholas Anthony worked on.

Daniel Edwards received the bachelor's degree in physics from the University of California, Irvine, in 1960. He has extensive experience in transducer design, including piezoceramic transducers for undersea mapping and for geological surveying. He has been with the Pellston Corporation since graduation. From 1960 to 1975, he was with the Music Division, where he carried out acoustic research studies and designed a series of speakers for Pellston amplifiers. In 1975, he joined the Electronics Division to direct Pellston's efforts in transducer design.

Colin Andrews joined Pellston just a month ago. He has a doctorate in electronics from the University of Toronto (1975). His thesis was titled Satellite Telemetry and its Applications to Geological Research. He obtained bachelor's (1964) and master's (1966) degrees in electronics from Duke University. Since 1975, he has worked for a "think tank" called DeLaire Research in Vancouver. During that time, he published 43 papers on telemetry systems. He is an expert in low-power transmitter design. Between 1966 and 1972 (when he went back to school to get the doctorate), he worked for Canco Oil in Calgary, designing signal processing equipment for seismic exploration.

11-3. Writing the Technical Sections for a Proposal and Assembling the Final Document

Although a proposal in response to a procurement request of this size in the real world would be far more complex than a correct response to this case, all of the elements and problems associated with proposal writing are present.

1. Carefully study the information in the following box, and outline it in a logical fashion. Write first drafts for the technical sections of the proposal you planned in Exercise 11-1 in accordance with your outline. Plan the illustrations for these sections at this stage. Make sure that each illustration has a figure number and a suitable title.

Note: Use the technical illustrations provided. Be sure to place figure numbers in the text just before the spaces in which illustration will be placed. Indicate a figure number and a suitable caption for each illustration. Pictures of the environmental equipment are not included, but you may include sketches of your own as an option.

2. When writing the final draft, rearrange the material as necessary for a coherent picture of the company's capabilities. Beware of the divergent styles. For example, one of your contributors writes only simple sentences. Another is careless with grammatical structures and punctuation, and so on.

3. **Special Instruction:** Although all of the technical material is supplied to you as first drafts written by contributing technical people, avoid "lifting" the expressions used by these contributors. Attempt to understand the situation, and rewrite the material to reflect a professional quality.

4. Go back through all of the sections to see if you have used the most suitable methods of exposition in all cases. Can you think of an example, a comparison, a graphic illustration, or an analogy that would improve readability? Also, double-check the grammar, spelling, and punctuation at this stage. Look for non-parallel structures, dangling modifiers, and poor subordination.

Note: All block diagrams, line drawings, etc., are illustrations. In other words, the final proposal will contain one sequence of figure numbers, not one for block diagrams and another for line drawings.

5. Prepare a title page, and write the introduction, the abstract, the table of contents, and the list of illustrations. Be certain to cite the correct page numbers in both the table of contents and in the list of illustrations. Make certain that no figure number or page number is larger than any of those

that succeed it in the list. (If such a situation does exist, you have made an error in numbering the illustrations.)

Note: As you will see in the contributions below, some of the team members habitually use non-metric units of measure. Be sure to convert all such measurements to the appropriate metric units.

The following contributions by each of the proposal team members were submitted to the group leader in accordance with the schedule:

From Nicholas Anthony

The temperature extremes and wind conditions mean pretty much that we have to use a metal framework that can be anchored into the permafrost. The other thing though is that the whole thing can't be too heavy so we have decided on an outer shell made of fiberglass. Protection of the instruments from the wind is of paramount importance so we will go to an aerodynamic design and that will make for an overall stronger structure windwise (if it doesn't fly away, ha, ha.).

Here is a diagram of the design.

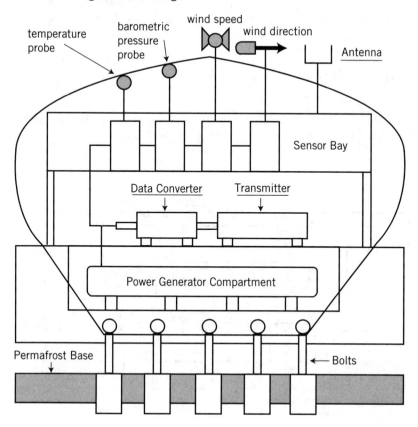

The metal framework is the only heavy part. The sensor bay and the power plant section are all made of glass fibers. The structure itself, that is without the power generator, the sensors or anything, calculates out to just under 260 kg. That includes the hardware to anchor it to the permafrost. I figure that the aerodynamic shape of the fiberglass bubble will prevent the thing from ripping apart. The only alternative would have been to have made the whole thing out of steel, and that would have doubled the weight because molding the steel into an aerodynamic shape would not be practical.

The dimensions of the base are four yards by four yards. We decided to use AN Type hardware because the buffeting that the thing might encounter could wear out ordinary stove bolts. The anchors will be drilled into the permafrost, and one-inch bolts will be fastened through the structure and into the anchors as shown in the diagram. There will be ten anchors, five on each side (the diagram shows only the five on the near side).

Some ground prep will be required for leveling. Minor adjustments can be made by loosening some anchor bolts and tightening others.

The power generating equipment will rest on the metal framework just above the tops of the anchor bolts. The sensor bay and the shelf that the data converter and transmitter sit on are all made of fiberglass panels, as of course is the outer shell.

Hope this is what you need. Good luck.

From Jean Wilberforce

Rita and I have decided on a combination of fuel cells and batteries with a backup gasoline-powered generator. The layout looks like this:

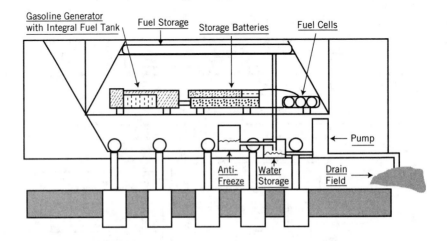

The fuel cells will supply power to the batteries. The batteries will supply power to the data converter, transmitter, and sensing units. The batteries are regular maintenance-free batteries. They are the same kind that car manufacturers put in new cars. These batteries have an excellent record. Some go 10 years without needing replacement.

The fuel cells will generate heat and water, besides electricity. The heat is good. The water is bad. A water storage tank has been included to collect the water. An anti-freeze tank has been included to mix with the water so it won't freeze. A pump takes the water and anti-freeze mixture out of the weather station into a drain field. You can see this on the diagram.

The backup generator will be driven by a three horsepower gasoline engine. It will start automatically when the battery charge gets below 18 volts. It will cut off when the charge is back up to 24 volts. The tank will hold enough gasoline for approximately 45 charges. The station can run for approximately 18 hours on one charge. Fuel cell failure will be signaled by a special transmission coded into the regular weather data. The unit will have to be repaired within 33 days of fuel cell failure. The station will shut down after 33 days.

The fuel cells and the batteries will generate a small amount of heat. The heat will keep the electronics gear at a temperature of –10 degrees F even if the outside air temperature drops to –80 degrees F. The amount of heat will be small enough to not cause a problem during the summer months.

From Rita Waechter

I understand that Jean has given you the information on the power plant. If you have any questions, be sure to see me.

I think it would be a good idea to include the project diagram along with the biographies of the project people. You might also mention that the project will have full access to the CAD (computer assisted design) facilities here at Pellston. All of the aspects of the design will be thoroughly computer tested before we begin building the prototype. All of the project members are completely familiar with CAD techniques, and we all are part of a LAN (local area network — interconnected personal computers). Also, we have software to support both the mechanical and the electronics design efforts.

From Daniel Edwards

Temperature will be measured by a metallic resistance device. It is the only type of thermometer that is tough enough to take the operating environment. The readout will be directed to a circuit containing a Wheatstone bridge and an automatic balance feature. The digital conversion system is included within the data converter unit.

An aneroid barometer will be used to monitor barometric pressure. This instrument will provide a voltage to the sensor bay component, which is essentially an amplifier. As with the temperature data, conversion to a digital signal is accomplished within the data converter unit.

Wind direction will be measured by a standard weather vane. No compass correlation is necessary, since the weather station is anchored in a known fixed position with regard to magnetic north and true north. The signal will be amplified and directed to the data converter unit for digital conversion.

A three-cup anemometer will be used to measure wind speed. Weather-resistant lubricants will be sealed within the shaft, a feature which should provide assurance that the shaft is free even at −80 degrees F. Tests in a simulated Arctic environment will be run on all of the sensing equipment before we install any of it on the prototype unit for final evaluation. The wind speed signal will be amplified in the sensor bay for transmission to the data converter for digital conversion.

The data conversion unit will process and encode all of the signals individually. Its coded output will modulate the transmitter, providing continuous data for transmission to the staffed stations, where the data will be decoded and printed out for incorporation into the six-hour Arctic weather forecasts.

The data conversion unit is also hooked up to the gas-powered generator. If the gasoline engine starts up, the data converter will add a special tag line to the weather data, telling the people at the other end that the fuel cells are malfunctioning. That is a signal for immediate maintenance.

A block diagram of the sensor system follows:

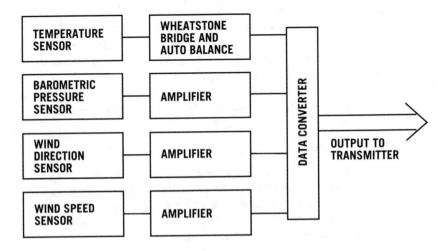

From Colin Andrews

The transmitter design is straightforward and uncomplicated. It is composed of a crystal oscillator, two doubler stages, and a power amplifier stage. The RFP requires a complete data broadcast every seven minutes. To satisfy this requirement, we will place a timing circuit in parallel with the power amplifier stage. Since each data transmission lasts for 14 seconds, the timer will activate the output for the first 14 seconds of each 7-minute period. The output will be to a quarter-wave dipole antenna.

Power to the antenna will be 12 watts, two watts greater than the specified power requirement. Here is the block diagram:

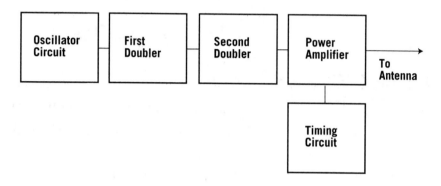

CHAPTER 12

Instructions

OBJECTIVES: In this chapter, you will discover that the instruction category embraces several kinds of documents. You will learn that manufacturing processes must be delineated so that shop personnel know exactly what procedures to follow in making the product. You will also learn that test procedures that help to assure product quality need to be written and that operating and maintenance procedures need to be provided to users of the product. And, most important, you will learn how to write instructions.

The Principles of Instructional Writing

Whether you are writing instructions to strike a match, to change a tire, to prepare a chemical formula, or to put a satellite into orbit, the guidelines presented in this chapter will help you produce a successful document. The first step, as with a proposal or a report, is to determine your purpose and to learn something about your audience.

What Is the Purpose of the Instruction?

Begin by considering the final result you expect those following your instructions to achieve, and make a topic (heading) outline as an initial plan. If you were writing a test procedure, for example, your outline might include some or all of the following headings:

1. Purpose of Test

2. Test Equipment Required

3. Test Setup

4. Test Procedure

5. Observation of Results

6. Action Required

Instructions can have more than one purpose. A maintenance manual, for example, may direct the user to perform certain tasks necessary for the upkeep of equipment. Its secondary purpose, however, may be to help the technicians performing the maintenance learn about the theory of operation so that they may diagnose and solve unanticipated problems. For this reason, maintenance manuals often contain theoretical discussions and other information of potential interest.

Who Will Carry Out the Instructions?

Knowing the audience for instructional writing increases the likelihood of success. Will your instructions be carried out by trained technicians? What will be the lowest level of reading ability of those who will use the instructions? Answering these questions will help you answer the specific usage and structural questions that will occur to you as you write the document.

The Gunning Fog Index, mentioned in Chapter 5 and contained as a feature of some word processing applications, can be of particular help to you as a writer of instructions. The Fog Index number roughly correlates to the grade level a reader will need to have completed to understand the writing. If the typical user of the manual has little or no post-secondary education, you will want to keep the Fog Index below 12, preferably at 8 or 10. Incidentally, you may be interested to know that *Time Magazine* has a fairly consistent Fog Index of around 10, although most of its readers have an education level of high school graduation or beyond. Even the *Atlantic Monthly,* which is read mostly by post-secondary graduates, tests between 12 and 14.

Choosing An Appropriate Writing Style

Illustrations in Instructional Writing

Specifically addressed in Chapter 8, illustrations are essential in technical writing. Nowhere is the value of good illustrations more apparent than in instructional writing. Some relatively simple instructions can be followed by reference to illustrations alone.

If you have ever attempted to put together a newly purchased child's toy, a lawnmower, or other device that was delivered in a disassembled state, you already know how valuable illustrations can be to those attempting to carry out instructions.

Equipment Descriptions and Theories of Operation

The appropriate writing style within these sections of an instructional document is similar to that used in proposals and reports. You will make generous use of illustrations, such as block diagrams, schematics, and exploded-view drawings *(spatial mode of exposition)*. And you will use a *linear mode of exposition* throughout these sections of the document, carefully choosing from among the four *reader-centered methods of exposition*. You will, of course, cast most sentences in a third-person, indicative-mood structure and arrange long discussions in a *top-down organizational pattern*.

Step-by-Step Procedures

In these sections of instructional documents, you will follow a unique approach. You will number (or designate as directed by specification) all the steps you expect the user to follow, and you will use second-person, imperative-mood structures for all such designated steps. The following example of a series of steps illustrates this style and tells how to write instructional steps:

1. Number all instructional steps, and write all such steps in second-person, imperative-mood structures.

2. Write a complete sentence for each step; do not omit words such as *a, the,* and *some,* unless the specification you are following tells you to do so. Each numbered step may also include information that specifies why the step is necessary. This information need not be in second-person, imperative-mood. However, you must avoid making the steps too long, or you will destroy the continuity between them.

3 If necessary, interrupt the series of steps to insert information that cannot easily be handled as part of a step.

4. Include note, caution, or warning statements wherever they are needed. **Notes, Cautions,** and **Warnings** are standard ways of communicating special information. Each has a specific meaning:

   A **Note** indicates that the information, although it is unessential for the immediate procedure, is of interest and potential use; it is thus written in the third-person, indicative mood.

   A **Caution** signals that equipment damage can result if the specific steps given in the caution statement are not followed. It is written in the second-person, imperative mood.

   A **Warning** is the most serious of the three, signifying that physical injury to personnel can result from failure to follow the instructions. *Cautions and warnings must always be written in the second-person, imperative mood.*

The following example includes procedural steps written in the second-person, imperative mood. Notice also that two "Warnings" and a "Note" make up a part of the procedure. Numbers have been used to make the procedural steps stand out from the other items of information, and boldface type has been used to emphasize key words.

SPIN RECOVERY TECHNIQUE

As stated elsewhere in this handbook, the CC-1 Silver Streak[1] is not certified for spins, and intentional spins are prohibited under any circumstances. However, should the airplane enter an unintentional spin, carry out the following steps without delay.

1. Set the **ailerons** to **neutral** and the **throttle** to **idle**.

2. Determine whether the airplane is spinning right or left, apply **full opposite rudder,** and immediately push the **stick well forward**.

WARNING

Refer to the turn coordinator if disorientation prevents your picking up the direction of rotation by visual reference outside the airplane.

3. **Hold** these **control inputs** until the rotation stops and the stall is broken.

4 When the airplane has stopped spinning, **center** the **rudder pedals,** return the **stick** to **neutral,** and allow the airplane to recover from the dive.

WARNING

Do **not** pull the stick back too briskly or too far aft. Trying to force the airplane out of the dive prematurely can produce a secondary stall/spin.

NOTE

Several hundred variables affect the spin characteristics of every airplane. Even trainer aircraft that are certified and used in spin training can be unpredictable under certain conditions. Some experts hold that every airplane has a spin mode from which recovery is impossible.

As you can see from the foregoing, the intended audience is made up of trained pilots. The reader is expected to know what a stall is, what a spin is, what the rudder and ailerons are and what cockpit controls move them, what position the throttle is in when it is set at idle, and what a turn coordinator is. As you will see in the subsequent paragraphs, nearly all instructions assume some training on the part of the readership.

---

[1] The CC-1 Silver Streak is a fictional airplane, and the instructions for spin recovery are, therefore, also fictional.

Types of Instructions

Manufacturing Process Procedures

Every company has its own method for dealing with this subcategory of instructions. Some produce manuals of standard procedures, augmenting them with materials sheets and process sheets. At the other end of the spectrum, some companies rely heavily on the experience of supervisors and specialists in the various shop areas, providing them with a minimum of written instructions.

In either case, the principles of good instructional writing apply to the production of all manufacturing process procedures.

Quality Control and Test Procedures

All manufacturing firms of any size maintain clearly written quality control procedures. These documents include everything from sampling methods to final test procedures for the product.

Materials, components, and subassemblies need to be checked for quality and reliability before they are accepted from vendors. When such purchases are in very large volume, testing each piece may be impractical. Therefore, statistical sampling methods often need to be employed for assurance of reliability at an acceptable level of probability. The well-conceived quality control manual gives all the information necessary to establish sampling schedules for these items.

In addition, the quality control manual establishes test procedures and prescribes acceptable results for each product. As one may imagine, the job of keeping the quality control manual current can be a major one.

In general, test procedures are composed of at least four parts:

- purpose of test
- test setup
- test procedure
- assessment criteria

Assessment criteria is sometimes called results or observations and sometimes includes action to be taken. Since nearly the entire document is composed of direct instructions, most of it is written in the second-person, imperative mood. The test setup part often includes an illustration. The following is an example of a simple test procedure that does not need an illustration:

HARDNESS EVALUATION PROCEDURE FOR STRUCTURAL TUBING

Purpose

The purpose of this procedure is to determine that the microstructure of welded structural tubing in the fuselage section of the CC-1 Silver Streak is not weakened by the welding process employed in manufacture.

Test Setup

1. Use sampling procedure VS-22 of the Pellston Quality Control Manual.

2. Place the butt-welded assembly in the tensile strength test unit.

Test Procedure

1. Subject the sample to tensile stress until a fracture occurs.

2. Section and mount 5 cm of the tubing on both sides of the fractured weld.

3. Immerse the specimen for 10 seconds in a solution of 97% H_2O, 1.5% HNO_3, and 1.5% HF.

4. Microscopically examine the specimen at 10×, 100×, and 500×, and record the incidence of cracks in the tubing.

Assessment Criteria

1. Reject any sample in which cracking in any part of the longitudinal cross section is evident.

2. In the event that a sample is rejected, discontinue the sampling procedure described in VS-22, and test all units.

3. Continue the 100% test regimen until 25 units in succession are free of defects.

4. When 25 successive units have tested free of defects, resume the tests in accordance with VS-22.

It should be noted that test procedures are also sometimes required as line items of contracts, especially contracts for prototype equipment.

Installation, Operating, and Maintenance Instructions

Most people are familiar with this kind of instruction. When you buy a new VCR or some similar device, you receive a set of instructions for setting it up and using it. And when it requires repair, you take it to a maintenance shop at which repair manuals are available to the technicians who will repair it.

The length and complexity of the manuals telling how to use or repair a device are usually in direct proportion to the complexity of the device itself. The manuals to support a military aircraft can run to thousands of pages. However, VCR or jet fighter, the general principles that govern good instructions are the same: instructions must be clear, concise, and unambiguous.

Handbooks and manuals often provide information in addition to instructions, since they are often regarded as teaching tools as well as reference tools for particular operations. Let's look at a page from a typical (although fictional) handbook of maintenance instructions, and let's see how the writer has handled the problem of telling a technician how to do a certain job. Note the uniformity in the following example. This uniform characteristic would come about because the instruction would have been written in accordance with a specification. Specifications that tell how a particular handbook or manual is to be written are for the purpose of standardizing publications, and as such they are an aid to clear communication. Occasionally, such specifications cover a great deal of ground, and they may dictate the inclusion of material that seems extraneous. Nevertheless, writers are better off with specifications than they are without them.

7-221. **GUNS AND RELATED EQUIPMENT**

7-222. **DESCRIPTION.** The complement of armament equipment includes two 20 mm, electrically operated machine guns. Along with the circuitry associated with the control and operation of the guns, a gun charging system and bore sighting equipment are also included.

NOTE

The equipment herein described has been designed to the standards set forth in MIL-ARM-4352G. Therefore, C-level maintenance is not authorized.

7-223. **INSPECTION.** Inspection after each mission is recommended. Both guns must be inspected at this time. Both guns must be removed at 25-hour intervals for thorough cleaning and inspection.

7-224. **REMOVAL.** The procedure given in this paragraph is for removal of the port-side gun. The procedure for removal of the starboard-side gun is given in paragraph 7-225.

WARNING

Ensure that all external power is disconnected, that armament switch is in "off" position, and that area is secured.

1. Open gun bay access door (station y-233).

2. Press "RELEASE" on pneumatic charging mechanism.

3. Unload magazine.

CAUTION

Do not cycle gun feed mechanism while magazine is empty.

4. Rotate gun in counterclockwise direction (viewed from muzzle end).

Note that the author has used third-person, indicative mood and third-person, subjunctive mood in the descriptive paragraphs, but has stayed exclusively with the second-person, imperative mood in all procedural steps.

Did you also notice that indefinite and definite articles (*a, an,* and *the*) are omitted from all second-person, imperative-mood statements? This omission would have been in response to a specification requirement. You should know that all specifications do not require the omission of articles, and that if you are not specifically directed to omit them, you should include them. Many people argue (quite reasonably) that the omission of the article is a distraction and that *Open the gun bay access door* sounds better and is more easily understood than *Open gun bay access door.*

The Software User's Manual — a Special Case

The popularity of computers has spawned a new industry: software user's manuals. One of the differences between these manuals and many other instructional documents is the lay audience to which they are quite often (although not always) directed. While a writer might expect a television repair manual to be used by a person with specialized training or a pilot's operating handbook to be used by a trained pilot, the writer of a software user's manual often faces the prospect of a largely untrained readership.

Another difference is the form of the final product, which can be a published book, on-screen help, or often both.

TYPICAL PHASES OF A SOFTWARE DOCUMENTATION PROJECT

The Research Phase

The first phase of a software documentation project is just like any other technical writing job, and the writer must ask all the same questions:

1. If there is a contract, what does it specify?

2. Is there a governing specification or style guide?

3. Who will be the technical liaison person?

4. What sort of communication skills does the technical liaison person have?

5. If the writer is not an in-house technical writer (some manuals are written, under contract, by outside firms), what form (interview, mail, e-mail, telephone, etc.) will the writer use to communicate with the technical liaison person?

6. What materials will the writer be working from?

7. Are there previously completed software manuals to serve as guidelines?

After answering the foregoing questions, you as the writer must answer some technical questions by consulting documents or by communicating with your technical liaison person.

1. How big is the application?

2. How many commands will you have to deal with and explain?

3. How stable is the design?

4. Will there likely be last-minute changes?

5. Is there a final "freeze" date, after which you can expect to proceed with the copyediting and production phases of the project?

If you are an outside contractor, the next step is to try to arrange for a demonstration, so that you can get a firm idea of the scope of the job. With a good idea of the number of pages and screens you will be expected to produce, you can decide on the best method of production: electronic layout or manual paste-up.

The Writing Phase

This phase should follow the writing process described in Chapter 6. A reasonable estimate for a first draft of a user manual is three to five hours per page. The rewriting stage, which includes developmental editing and copyediting, can easily add another hour for each page of final copy.

The following sample[2] is courtesy of Shana Corporation, developers of Macintosh and Windows software applications. This particular sample is from the documentation that Shana Corporation provides with its electronic forms design application, Informed Designer©.

Combs

Combs are lines that divide the character spacing of fields and tables into equally spaced sections. Use the Combs command to specify the number and height of combs in a field or table cell.

*To add combs, first select the cell in the field or table, then choose **Combs...** from the Settings menu. The Comb Settings dialog box appears:*

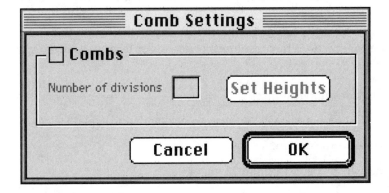

Click the 'Combs' checkbox to turn comb lines on or off. When combs are on, you can enter the number of divisions in the text box provided. Furthermore, you can adjust the height of each comb line individually. To do this, click the 'Set Heights' button. The Comb Heights dialog box appears:

² From Shana Corporation's official documentation. Used with permission of Shana Corporation, Edmonton, Alberta.

Using this dialog box, you can click and select each comb divider in the scrolling list, then enter a different value in the text box below the list. To change more than one divider line at a time, select each one while holding down the Control (Windows) or Command (Mac OS) key, then enter the new value.

You can measure comb heights in either percentage of cell height, or points. If you use the percentage of cell height option, the comb heights will automatically adjust when you resize the field's cell. Set the comb height by clicking either the '%' or 'Points' radio button, then enter a value in the text box. Click 'OK' when you're finished changing the comb heights. To cancel the Combs command, click 'Cancel.'

Checkboxes

Fields and table columns that are configured as checkboxes display a check symbol instead of a textual value. Use the Checkbox command to configure a field or a table column as a checkbox.

To configure a checkbox, select the field or table column that you want to change, then choose **Checkbox...** from the Settings menu. The Checkbox Settings dialog box appears.

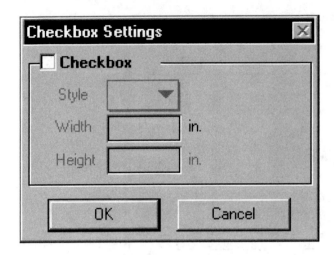

Click the 'Checkbox' option to turn the checkbox setting on or off. After you turn the checkbox option on, choose a checkbox style from the 'Style' drop-down list. Specify the dimensions of the checkbox by typing values in the 'Height' and 'Width' text boxes. If you enter a dimension that exceeds the size of the cell that holds the checkbox, Informed Designer will alert you with a message.

As a shortcut to the Checkbox command, you can also select a field or table column, and then click the checkbox button on the Cell palette.

Checkbox button

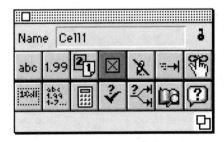

The selected field will display a checkbox with the default style and size;
that is, a plain frame measuring 0.1667 inches by 0.1667 inches, with
an 'X' check symbol. For more information about the Cell palette, see
"Using the Cell Palette" in Chapter 1 of your Informed Designer Forms
Automation *manual.*

Exercises

Note: No special technical knowledge is required to complete the exercises in
this chapter.

12-1. Writing a Customer Service Bulletin

1. Carefully read the following case study.

2. Since this instruction will be mailed to those who carry out the instructions
   and since the instruction is fairly short, use a business letter format.
   Address the letter to all owners of the ELT-7C, manufactured by Pellston
   Electronics. Make up a letterhead similar to those found in the samples of
   Pellston letters elsewhere in the book.

3. Outline the letter, keeping in mind that your purpose is to motivate owners to use the equipment properly.

4. Write the first draft, calling attention to the problem; be sure to cite the statistics.

5. Write a final draft of the letter, paying special attention to word choices that
   might motivate the owners to action. Pay special attention to tone. An authoritative tone that does not condescend would be best for maximum results. Use the illustration in the instructional part of the letter.

6. Incorporate the illustration into the letter, converting the dimensions given
   on the drawing to metric units.

7. Do a final check for grammar, punctuation, spelling, and clarity of
   expression.

You have just begun work at Pellston when Mr. Dobbs telephones you to say that a problem has occurred in the Electronics Division. He also says that you are being temporarily assigned to look after the problem and that your job title is Special Assistant to the President. He begins, "Last year, we got into the ELT business. Do you know what an ELT is?"

"Yes, I'm familiar with our product," you answer. "The ELT Model 7C is, I understand, already very popular among owners of small private airplanes, mostly because of its low cost."

"Right. Its function is to provide an FM radio signal to search aircraft in the event of a crash or a forced landing in a remote area, and its presence is required by law on any aircraft that is flown more than 25 miles from its home airport. When activated, the ELT Model 7C operates simultaneously on two frequencies: 121.5 megahertz in the VHF range and 243 megahertz in the UHF range. It is mounted on a bracket in the airplane and is positioned in such a way that a shock load of 5 g or more will cause it to begin transmitting a signal, provided that the 3-position switch is in the 'standby' position.

"This 3-position switch, the device's only control, may also be placed in the 'on' position or in the 'off' position. The purpose of the 'off' position is to provide the user a means to conserve the battery's power while awaiting rescue for a prolonged period of time. The downed pilot may then position the switch to the 'on' state when a search aircraft shows up in the area.

"Another feature of the ELT-7C is its portability. It may be disconnected from the antenna on top of the aircraft's fuselage and operated through its own built-in antenna. This feature is important for two reasons: If the aircraft should come to rest in an upside-down position, the aircraft antenna might become damaged or ineffective; also, the downed pilot may elect to abandon the aircraft and attempt to 'walk out,' carrying the ELT-7C along.

"A problem has come to light, however, since the ELT-7C's introduction on the market. Many pilots place the 'on-off-standby' switch in the 'off' position in order to prevent the unintentional activation of the device that can occur when the airplane touches down hard during a normal landing. Of course, if the device is off when a plane is forced down, the pilot can simply turn it on."

"If physically able to do so," you reply.

"Exactly. Last year, three pilots were not able to switch it on; two of them might have been saved had they been found sooner. We need to have a customer service bulletin sent out in the form of a letter, explaining the situation to pilots, and telling them to put the ELT on their pre-takeoff checklists. Make it informative and forceful."

Knowing the value of technical illustration, you decide to include a diagram of the ELT in the service bulletin. You search the files for a suitable photograph or drawing of the device. Here is what you find.

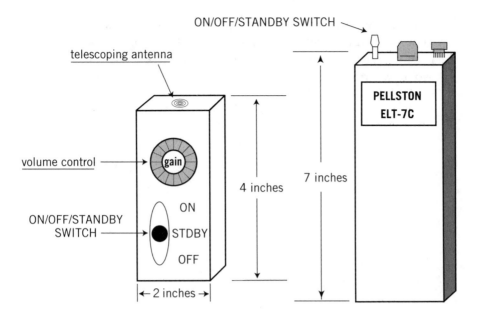

ON/OFF/STANDBY SWITCH

telescoping antenna

volume control — gain

ON/OFF/STANDBY
SWITCH

ON
STDBY
OFF

4 inches

2 inches

7 inches

PELLSTON
ELT-7C

12-2. *Writing an Instructional Memorandum*

1. Review the boxed material given as part of Exercise 9-3 in Chapter 9.

2. Write the memorandum to the dealers. Complete all the steps, from out-
    lining to final reading, given in Chapter 6. Make sure that you emphasize
    the positive aspects of the new warranty policy. Also, you will have to tell
    the dealers that warehouse personnel will pick up any instrument that is
    returned for repair under warranty and that the warehouse will handle
    all packaging and shipping. The dealer's responsibility will be limited to re-
    ceiving the instrument from the customer and recording the complaint
    as clearly as possible. You can design a form to be used by dealers in han-
    dling the complaint, and you can include the form as an attachment to
    your memorandum to them.

3. Write a memorandum to warehouse personnel, including a set of directions
    for handling warranty work under the new policy. Complete all the steps,
    from outlining to final reading, given in Chapter 6. Be certain to use the
    second-person-imperative for all steps that you want the warehouse per-
    sonnel to take in carrying out the policy.

4. Write the new warranty booklet. Organize the instructions for "Caring for Your New Guitar" with plenty of headings. Be certain to use the second-person-imperative for all steps that you want the new owner to take to avoid unwarranted damage.

CHAPTER 13

Correspondence

OBJECTIVES: An ever-increasing number of technical and administrative workers handle their correspondence without the help of clerks or secretaries. In this chapter, you will learn about two widely accepted letter formats and some of the reasons for adhering to them. You will also learn of the various kinds of letters (both internal and external) you might be expected to write, the importance of carefully framing your letters, and ways of achieving a proper tone.

Correspondence and the Technical Writer

Correspondence forms are common in technical writing. For example, as stated in earlier chapters, external reports and proposals are often written in letter format, and most internal reports, proposals, and instructions are written in memorandum format. Beyond these uses of letter and memorandum formats, a broad spectrum of letter categories grows out of the vendor/customer relationship. Although each letter within this spectrum could be loosely classified as a report, proposal, or instruction, some of them fit better into another set of categories, namely: claims, adjustment, collection, request, and request responses. Subordinate to these categories, good-news and bad-news letters have distinguishing features, as well.

Except for those letters sent as faxes or electronic mail, business letters are sent through the mail or other external courier services. Memorandums remain within a company's internal mail system, whether the system is electronic or traditional.

Letter Style and Tone

Your letter-writing style says a lot about you. A direct, no-nonsense style tells the reader that you are businesslike and efficient. However, a style that is too direct or terse might indicate impatience or even a disdain for the reader. A flowery, indirect style may suggest to some readers that you have something to hide or that you are a bungler, a time-waster, or a throwback to a previous era. Balancing styles is a matter of finding something called *proper tone*. That is, the words you put on paper make a "sound" as they echo inside a reader's head, just as the words you speak make a sound that strikes a listener's eardrums.

The differences in speaking and writing come about because a writer can neither soften harsh words by speaking softly and looking sympathetic, nor emphasize soft words by speaking harshly and jumping up and down. Thus, when you write, you must choose words and structures much more carefully than when you speak. That is, your words and structures must convey the tone you intend without the benefit of your voice and demeanor. And everything you write has tone whether you are conscious of it or not.

The tone of a letter reveals the attitude and general state of mind of the writer at the time of writing, and mature writers make every effort to control tendencies to anger and impatience. Good letter writers analyze situations carefully, considering all the possible motives of others and measuring their own responses.

We will see illustrations of these points in our examination of specific letter categories. You should also note that the principles of courtesy and tone apply to all letters, whether they be memorandums or external letters. Neither does the method of conveyance, whether it be an internal mail system, Canada Post, courier, fax, or e-mail, remove the need for courtesy and proper tone.

Some Specific Letter Categories

Letters of Complaint and Letters Resolving Complaints

Since a complaint letter is a bad-news letter, tone is especially important. An angry letter written to register a legitimate complaint will not lead to a remedy as quickly or as certainly as will a measured, objective description of the difficulty.

A letter responding to a complaint may fit the good-news or bad-news subcategory, and a person responding to a complaint must, first of all, determine which kind of a letter to write in reply. Understanding the complaint is, therefore, extremely important. For the purpose of showing the importance of a mature style and tone in both types of letter, we will begin with an unjustified complaint letter and follow the development of the correspondence. As you will see in the exchange of letters that follows, a company representative who answers complaint letters by merely stating company policy can do a lot of harm to the company.

321 Rosedale Lane
Thunder Bay ON P7P 3T7
December 15, 1998

Pellston Corporation
Musical Instrument Division
567 Cherry Blossom Avenue
Surrey BC V4D 3Y7

Dear Sir or Madam:

One of your instruments has failed because of faulty workmanship and materials. When my son slipped on an icy sidewalk while carrying his bass viol, the instrument was smashed beyond repair. I warned him about the ice on the sidewalks that day, but he didn't listen. He never listens.

Fortunately, he was not seriously injured, since the collapsing instrument cushioned his fall. Nevertheless, I am out the cost of a bass viol. Don't you agree that some token compensation might be in order?

Thank you.

Yours truly,

Vera Maelstrom

Vera T. Maelstrom (Mrs.)

A careful reading of this letter reveals a good deal about the writer and about the situation she is describing. First, there is a hint of anger in this letter, even though the writer has been careful to control herself for the most part. The comment about her son not heeding her warning gives the first indication that this letter might not be about faulty workmanship and materials at all. Is she really angry with Pellston Corporation because the bass viol broke when her son fell on it? Or is she perhaps seeking some kind of an outlet for the anger she feels toward her son, who failed to heed her warning? Of course, one could not be absolutely sure that this conclusion is correct. Nevertheless, the letter's overall tone strongly suggests that Mrs. Maelstrom just wants someone to understand her frustration and, maybe, offer a little sympathy in the form of token compensation.

Had the person responding to this letter been especially sympathetic and attentive to her problem, the exchange might have ended at that point. However, the customer relations manager reading this letter (as we will see in the response that follows) makes no attempt to analyze the situation. In addition, he makes two grave errors, one of which gives Mrs. Maelstrom something on which to vent her, as yet, unspent anger. See if you can spot these errors.

SAMPLE 13-2 | **A Badly Written Reply to a Letter of Complaint**

Pellston Music Division
567 Cherry Blossom Avenue
Surrey BC V4D 3Y7

Telephone: (604) 555-5000   **Fax: (604) 555-6000**

December 18, 1998

321 Rosedale Lane
Thunder Bay ON P7P 3T7

Dear Ms. Maelstrom:

We are in receipt of your request for compensation regarding the destruction of an instrument. Unfortunately, our replacement policy does not cover such an eventuality as you describe in your letter.

We are sorry we cannot comply with your request, but we hope to continue serving you with quality products. Thank you.

Yours truly,

Wallace Waffle

Wallace Waffle
Customer Relations

WW: ec

Mr. Waffle's reply is polite. He even uses the words *unfortunately* and *we are sorry*. Did you spot the two errors, though? First of all, Mrs. Maelstrom had requested only "token" compensation. Had Mr. Waffle read her letter carefully, he would have realized that she hadn't asked for a replacement of the instrument. He could have offered her the understanding she was probably seeking. A comment, such as "I know how kids are — my two never listen to a word I say," and a five-dollar gift certificate would likely have ended the matter. She might even have been inclined to ignore the other error, that of his failing to note that she had stated her preferred title to be *Mrs.*

But these two errors taken together seal Waffle's fate.

SAMPLE 13-3 | **An Angry and Ineffective Letter of Complaint**

321 Rosedale Lane
Thunder Bay ON P7P 3T7
December 21, 1998

Pellston Corporation
Musical Instrument Division
567 Cherry Blossom Avenue
Surrey BC V4D 3Y7

Dear Mr. Waffle:

Just who do you think you are, calling me Ms.? It's no wonder that you can't build a decent musical instrument. You can't even read! And then you have the gall to invite me to continue to buy the junk you put out. Do you think I'm an idiot?

You say that you don't cover "such an eventuality." Are you implying that my son is overweight?

You had better pull up your socks, junior, or face a lawsuit! I expect a new bass viol sent by air express, and I expect it tomorrow!

Yours truly,

Vera Maelstrom

Mrs. Vera T. Maelstrom

At this point, Mrs. Maelstrom has taken off the gloves. In her mind, her appeal for understanding has fallen on the ears of a lout. Now the pent-up anger flows out of her like acid from a broken laboratory jar. She doesn't really want a new bass viol; she wants an adversary, and like it or not, Waffle is it.

At this stage, Waffle needs desperately to understand that acceding to this customer's demand will end the situation, but that a second way out may still exist. An abject apology, an expression of understanding, and token compensation might yet salvage the relationship with this difficult customer. However, Waffle is human too (although Mrs. Maelstrom will not think so, her ability to empathize being even less than his), and so he falls back on company policy in a letter whose general tone is not particularly conciliatory. He closes with a holiday wish that serves only to highlight an overall impression of non-interest and detachment, in effect pointing away from this problem she finds so consuming.

SAMPLE 13-4 **A "Company Policy" Reply to a Letter of Complaint**

 Pellston Music Division

567 Cherry Blossom Avenue
Surrey BC V4D 3Y7

Telephone: (604) 555-5000   Fax: (604) 555-6000

December 24, 1998

321 Rosedale Lane
Thunder Bay ON P7P 3T7

Dear Mrs. Maelstrom:

My heartfelt apologies for referring to you as Ms. Maelstrom. I realize that some married women are offended by the title, and I regret having used it. We here at the company use Ms. on all of our letters to women. I hope you will understand.

Unfortunately, there is nothing I can do about the damaged instrument. When someone falls through a bass fiddle, whether he be fat or skinny, he has used the instrument in a way in which it was not intended for use.

continued ➤

May I take this opportunity to wish you and your family a Merry Christmas and a Happy New Year?

Yours truly,

Wallace Waffle

Wallace Waffle
Customer Relations

WW: ec

---

321 Rosedale Lane
Thunder Bay ON P7P 3T7
December 29, 1998

Pellston Corporation
Musical Instrument Division
567 Cherry Blossom Avenue
Surrey BC V4D 3Y7

Mr. Waffle:

What do you mean some married women are offended by the Ms. title? All the women I know are offended by it. I understand all right, Waffle. I understand that you are an insensitive jerk, who works for a company that employs managers who are looking for a power trip — a bunch of half-wits pretending to be big executives.

All you seem to be able to do is to cite company policy. Are you really alive, Waffle? Or are you some kind of a machine? I'll bet you're a computer, aren't you, Waffle? No human being is named Wallace Waffle.

Anyway, machine or human, I'm sick of dealing with you, Waffle. You had better tell your supervisor that Mrs. Maelstrom is losing her patience.

Yours truly,

Vera Maelstrom

Mrs. Vera T. Maelstrom

As the letter indicates, Mrs. Maelstrom seems to be just getting warmed up. She is eager to continue the fight, and the identity of her opponent doesn't concern her much. This letter makes it clear Mrs. Maelstrom is the kind of person who will take advantage of a perceived weakness. She believes she has defeated Waffle in their verbal battle, and now she is spoiling for combat with someone higher up.

Unfortunately, her next match proves to be with someone as cantankerous as she is. Mr. Peabody, the general manager, does not have time for games. In his view, if it's conflict she wants, it's conflict she'll get.

SAMPLE 13-6 | **A Completely Inappropriate and Non-Businesslike Reply to a Letter of Complaint**

Pellston Music Division

567 Cherry Blossom Avenue
Surrey BC V4D 3Y7

Telephone: (604) 555-5000

Fax: (604) 555-6000

January 3, 1999

321 Rosedale Lane
Thunder Bay ON P7P 3T7

Ms. Maelstrom:

We aren't about to replace an instrument that some fat person fell on. Do you think we are idiots?

You want to sue, Maelstrom? Be my guest!

Yours truly,

Harry Peabody
Vice President and General Manager

HP:bv

This situation is obviously going to have to be sorted out by someone else. Both Mrs. Maelstrom and Mr. Peabody are angry people who don't consider the outcome of their actions. Writing an angry letter might be good therapy (ridding one of the frustrations of daily life), but it is bad business practice.

Mrs. Maelstrom, like most bullies whose bluff gets called, is subdued by Mr. Peabody's outburst. Her final letter is an I'm-telling-on-you letter.

SAMPLE 13-7   An I'm-Telling-on-You Letter of Complaint

321 Rosedale Lane
Thunder Bay ON P7P 3T7
January 7, 1999

Grant Dobbs, President
Pellston Corporation
Musical Instrument Division
567 Cherry Blossom Avenue
Surrey BC V4D 3Y7

Dear Mr. Dobbs:

I herewith enclose copies of the correspondence I have had with your Musical Instrument Division. I can't believe that you wish customers of the Pellston Corporation to be treated in this way. Do you really want to be sued, Mr. Dobbs?

Yours truly,

Vera Maelstrom

Mrs. Vera T. Maelstrom

Letters of Request

Letters asking for information or ordering goods and services fall into the letters-of-request category. Request letters are very easy to write. They usually consist of three paragraphs, the first indicating the nature of the request, the second citing the specifics of the request, and the third thanking the addressee. The style should be straightforward and without much preamble, producing a tone that is respectful without being wasteful of the addressee's time.

Notice that Sample 13-8 opens with a brief description of the reason for and the nature of the request. It then asks for specific information about the product and closes with a *thank you.*

The Good-News Reply

If you have good news, relate it in the first sentence of the letter and provide necessary details afterwards. It is pretty hard to write a bad good-news letter. Those receiving the good news will tend to focus on the substance of the message, whereas recipients of bad news often focus on the letter itself, looking for something to shore up a collapsing optimism.

Samples 13-9 and 13-10 show how well-written good-news letters are structured. The first is a specific response to the request for information in Sample 13-8.

SAMPLE 13-8 | **A Well-Written Request Letter**

S P A C E
Magazine

1241 Anaheim Boulevard
Cement City CA 97531
Telephone: (714) 555-7500
Fax: (714) 555-7501

September 12, 1998

Pellston Corporation
Musical Instrument Division
567 Cherry Blossom Avenue
Surrey BC V4D 3Y7

Attention: Ms. Alicia Von Nordstrom
         Controller

Dear Ms. Von Nordstrom:

I understand that until recently Pellston Music Division manufactured and sold several lines of musical instruments that contained Brazilian rosewood.

continued →

As a reporter for *Space Magazine,* I am investigating the economic fallout resulting from the withdrawal from the market of all those products containing Brazilian rosewood. Would your department be able to furnish me with the following information?

1. The total number of instruments withdrawn from the market
2. The cost of recalling stock
3. The difference between gross sales for 1997 and projected gross sales for 1998 attributable to lost sales of these items

I will be most grateful for this information and any other facts or figures you may have that will shed light on my investigation. Thank you.

Yours truly,

Shelley Gianopolous

Ms. Shelley Gianopolous
Investigative Reporter

SAMPLE 13-9 **A Good-News Reply to a Request Letter**

P*ellston* Music Division
567 Cherry Blossom Avenue
Surrey BC V4D 3Y7

Telephone: (604) 555-5000   **Fax: (604) 555-6000**

September 18, 1998

Space Magazine
1241 Anaheim Boulevard
Cement City CA 97531

Attention: Ms. Shelley Gianopolous
            Investigative Reporter

continued ➤

Dear Ms. Gianopolous:

Pellston Music Division is pleased to comply with your request for information regarding Pellston's losses ensuing from the ban on Brazilian rosewood products.

In compliance with the new legislation, 834 instruments with a wholesale value of $582,000 were withdrawn from the market in January of 1997.

The difference in gross sales between 1997 and those projected for 1998, attributable to lost sales of the affected product lines, is estimated at $3,438,000.

Additional information that you may find useful might be that the closure of the BR-21/22 Guitar and the BR11/12 Mandolin production lines reduced Pellston's manufacturing work force by a total of 95 crafts workers and technicians, a figure representing 38 percent of the manufacturing total. Layoffs in staff departments such as accounting, purchasing, warehousing, shipping, and sales will take place within the next four months. Approximately 185 jobs will have been abolished by the end of the year.

I hope you will find this information useful. We at Pellston look forward to reading your article when it appears.

Yours truly,

Alicia Von Nordstrom

Pellston Corporation
Music Division
Ms. Alicia Von Nordstrom
Controller

AVN/ist

Pellston Corporation

4567 Airport Road
Sunnyside ON K2V 1M8

Telephone: (613) 555-5000
Fax: (613) 555-5333

An Industry Leader Since 1958

March 5, 1998

Mr. Matthew Galsworthy
654 Oliver Avenue
Starbutton ON L5A 7Y9

Dear Mr. Galsworthy:

We are pleased to offer you the position of Senior Technician in Pellston's Engineering Department. Details of the salary and benefits package offered are contained in an appendix to this letter.

Would you please respond to this offer on or before March 15, 1998. We look forward to working with you and to a long and mutually beneficial relationship.

Yours truly,

THE PELLSTON CORPORATION
Ms. Patricia V. Pilsnipper
Personnel Manager

Attachment
PVP/esp

The Bad-News Reply

Tone is important in all letters, but perhaps its importance can be seen most clearly in letters that must convey unpleasant facts. At times, the specific information, product, or service requested cannot be supplied for one or another reason. When you cannot completely honor a request, you should begin with a paragraph that includes a restatement of the request. Beginning in this way can clear up misunderstandings caused by letters of request that fail to clearly indicate what the writer wants.

The second paragraph should give the reasons why you cannot comply with the request, and the third paragraph should supply whatever relevant information is available.

An expression of the writer's hope that the information, etc., will prove useful provides an easy exit from the letter.

SAMPLE 13-11 | **A Bad-News Reply to a Request Letter**

*P*ellston Music Division

567 Cherry Blossom Avenue
Surrey BC V4D 3Y7

Telephone: (604) 555-5000   **Fax: (604) 555-6000**

September 18, 1998

Space Magazine
1241 Anaheim Boulevard
Cement City CA 97531

Attention: Ms. Shelley Gianopolous
Investigative Reporter

Dear Ms. Gianopolous:

Thank you for you interest in the economic difficulties suffered by Pellston Music Division as a result of the prohibition on the sale of any product containing Brazilian rosewood. As I understand your request, you are seeking specific information regarding direct financial losses.

Unfortunately, the ban on Brazilian rosewood products has forced operational changes producing a financial condition that cannot be defined by a mere calculation of the initial losses. While the number of guitars and mandolins recalled is relatively small, the dollar value of the Brazilian rosewood instruments we manufacture and sell accounts for 75 percent of our total sales each year. In view of this fact, you can no doubt see that an accurate as-

continued →

sessment of the loss must eventually include the dissolution of the Pellston Music Division.

Should the legislation that has led to this situation not be repealed within the next 30 days, the Division will be forced to close before the end of the year.

I hope you will find this information useful.

Yours truly,

Alicia Van Nordstrom

Pellston Corporation
Music Division
Ms. Alicia Von Nordstrom
Controller

AVN/ist

Another very common type of bad-news letter is the kind that must tell a job applicant that someone else got the job. Although strictly speaking, letters of this kind are not a direct response to a letter of request, they fall roughly into the same class. The following samples show the wrong and right ways to handle job refusal letters.

The facts given in the first example are direct and businesslike. The applicant will certainly know where he stands after reading this letter. But the tone is almost one of accusation and reproach. P.V. Pilsnipper displays a remarkable lack of feeling and tact in this letter.

Does directness of style really dictate the use of words such as *rejected* and *inadequate?* Did Pilsnipper intend to make Galsworthy feel like a fool for applying for this job in the first place? Perhaps Pilsnipper's intentions were merely to offer some useful advice, but the letter is no less reproachful than it would have been had it started: "Why are you wasting my time by applying for a job when you don't have the minimum qualifications, you nitwit?"

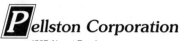

Pellston Corporation

4567 Airport Road
Sunnyside ON K2V 1M8

Telephone: (613) 555-5000
Fax: (613) 555-5333

An Industry Leader Since 1958

March 5, 1998

Mr. Matthew Galsworthy
654 Oliver Avenue
Starbutton ON L5A 7Y9

Dear Mr. Galsworthy:

Your application for the position of Senior Technician in our Engineering Department has been rejected. Your current level of preparation is inadequate for a Senior Technician's position, and I therefore suggest that you apply for a junior-level position.

Unfortunately, Pellston has no Junior Technician positions open at this time. We will keep your application on file for 30 days.

Yours truly,

Patricia V. Pilsnipper

THE PELLSTON CORPORATION
P.V. Pilsnipper
Personnel Manager

PVP/esp

Here is another version of the same letter.

Pellston Corporation

4567 Airport Road
Sunnyside ON K2V 1M8

Telephone: (613) 555-5000
Fax: (613) 555-5333

An Industry Leader Since 1958

March 5, 1998

Mr. Matthew Galsworthy
654 Oliver Avenue
Starbutton ON L5A 7Y9

Dear Mr. Galsworthy:

Thank you for applying for the position of Senior Technician in Pellston's Engineering Department. Because of the extensive supervisory responsibilities associated with this position, we have found it necessary to choose an applicant whose experience includes 12 years in the electromechanical field.

Nevertheless, we believe that the training you have obtained at the Northern Alberta Institute of Technology would qualify you for one of our junior positions. Unfortunately, we do not have any openings at the moment.

We routinely keep applications on file for 30 days, and we will review your application if an opening should occur during the coming month. If you do not hear from us within that time, we would appreciate your applying again. Thank you for your interest in Pellston Corporation.

Yours truly,

Patricia V. Pilsnipper

THE PELLSTON CORPORATION
Ms. Patricia V. Pilsnipper
Personnel Manager

PVP/esp

Some people would argue that there is little difference in the two letters, that the first few words tell this applicant he is not going to get the job, and that all the soft talk in the world can't change the facts. However, such a view of human relations is cynical, shortsighted, and blunt. It is a viewpoint that ignores human emotions.

The writer of the second letter is sensitive to the applicant's feelings. Moreover, she manages to avoid a fawning sympathy or insincere commiseration while at the same time giving the applicant the respect each of us deserves. How does the writer achieve this tone? She begins with something other than the bad news. Thanking Galsworthy for applying allows him to retain a bit of dignity when the inevitable bad news comes in the next sentence. Then she couches the bad news itself in positive terms (the many years of experience chalked up by his competitor) rather than in a blunt declaration that Galsworthy is an inadequate and, very possibly, a pitiful figure. The balance of the letter offers hope and continues to treat him with the dignity and respect owed him as a human being.

In summary, this letter consists of an opening sentence that acts as a buffer for the bad news to follow and of a positive finish that offers encouragement of some kind (depending on the nature of the bad news). Although you may find this formula useful, you will find that cultivating an attitude of real concern for the recipients of the bad-news letters you write is a much better approach to the problem.

Sales Letters

Direct marketing is an art unto itself. Therefore, sales letters ought to be composed by highly skilled specialists in direct-market advertising. In acknowledgment of this truth, we will not presume to say anything about sales letters, but rather to note that sales letters, along with brochures and other direct-market advertising, are *not* technical writing in the strictest sense of the phrase. Corporate communications, including such media as house organs and newsletters, although they are related to the world of work, fall generally into the journalism category and, likewise, lie outside the technical writer's scope of expertise.

Tactful Memorandums

Chapter 9 discusses the memorandum as it applies to the three main categories of technical writing. Like business letters, however, occasionally a memorandum does not readily fit the category of reports, proposals, or instructions. For example, one might have to write a memorandum to explain a policy and to tactfully request a change in another's method of doing a job. Although a memorandum written to fulfill this purpose might generally be classified as an instruction, the need to avoid giving offense would make the job of instruction more difficult than it would be otherwise, and a tactful tone would become an important quality.

Exercises

13-1. Satisfying an Irate Customer

As Grant Dobbs, review the series of sample letters between Mrs. Maelstrom and Wallace Waffle given in this chapter. Then respond to Sample 13-7. Be sure to perform all the steps in the writing process.

1. Write an outline for the letter to Mrs. Maelstrom, listing the major points to be made.

2. Select an acceptable format, either block or semi-block (refer to Chapter 9 if you need to).

3. Draft the letter.

4. Rewrite the draft, paying special attention to the previous exchange and keeping in mind that you are dealing with an extremely difficult person. The tone should be conciliatory and businesslike. Even though this person has reacted unreasonably, she has suffered real offense (Peabody's letter), and you should acknowledge her injury without making reference to her unreasonable attitude. After writing a first draft of the letter, follow all the rewriting substeps given in Chapter 6.

13-2. Writing a Letter Requesting Information

1. Write a letter to your local newspaper asking for information about the publication's editorial policy, hiring practices, or selection criteria for deciding what syndicated cartoon strips the paper carries.

2. Write a letter to a company or a government agency of your choice, inquiring about a subject of real interest to you. (You might choose to mail this one.)

13-3. *Writing a Letter Relating Good News*

1. Write a letter to the editor of your local newspaper agreeing with the viewpoint presented in an editorial or in a letter to the editor.

2. Write a letter to a manufacturer of a product that has served you well or to a restaurant at which you found the food and service to be of excellent quality.

13-4. *Writing a Letter Relating Bad News*

1. Write a letter to the editor of your local newspaper disagreeing with the viewpoint presented in an editorial or in a letter to the editor.

2. Write a letter to a manufacturer of a product that you found to be of inferior quality or to a firm that gave you poor service.

13-5. *Writing a Memorandum Reminding an Employee of the Need for Good Customer Relations*

1. Review the letters between Mrs. Maelstrom and the Pellston Corporation, which appear in this chapter.

2. Write a memorandum to Mr. Peabody, outlining the importance of good customer relations. Advise him against writing angry letters to customers in the future, at the same time emphasizing his value to the company.

3. Also, tell him of the disposition of the Maelstrom case (Exercise 13-1). The tone should be one of camaraderie, but not one that discredits Mrs. Maelstrom.

4. Be certain to follow all the steps in the writing process (see Chapter 6).

13-6. *Writing a Memorandum to Tell Employees of a Policy Change*

1. Write a memo to all Pellston divisions, outlining a policy for addressing letters.

2. When writing this memo, convince the employees who deal with customers of the desirability of courtesy.

3. You might also extend the discussion by pointing out that "stonewalling" by falling back on company policy is the opposite of being sensitive to a customer's needs.

4. Be certain to follow all the steps in the writing process (see Chapter 6).

CHAPTER 14

Oral
Presentation

OBJECTIVES: In this chapter, you will learn the eight principles of public speaking, the differences between speaking in an industrial situation and speaking on other occasions, and the methods for preparing and delivering an industrial speech.

The Principles of Public Speaking

Many books and articles have been written about public speaking, and most of them offer advice based on the theory of interpersonal communication in vogue at the time of writing. Some courses in public speaking require students to follow certain prescribed procedures, such as gesturing at important junctures in the speech or structuring the speech to conform to a formula. The principles that have endured, however, are those that could have been derived from common sense. They are as follows:

1. Be prepared.
2. Be confident and avoid affectation.
3. Use only Standard English.
4. Respect the audience.
5. Adjust your voice level to suit the environment.
6. Put variety in your voice.
7. Establish a smooth pace.
8. Maintain eye contact with the audience.

The negative effects of the violation of some of these principles are illustrated in the following account:

Two companies (which we will call Firm A and Firm B) were finalists in the quest for a large government contract. Both firms were asked to give an oral presentation of their respective proposals to a group of high-ranking government officials and contracts administrators during a visit to each firm's facility. At Firm A, the presentation was coordinated by the Technical Writing Department Manager, who introduced each of the proposal contributors in turn. Each of these contributors spoke, highlighting the major aspects of a part of the proposal. At the conclusion of the presentation, Firm A's vice president was to take the visitors on a tour of the plant.

All went well until the presentation was over, at which time the vice president, who was supposed to announce that he would now conduct the tour of the plant, for some reason felt obliged to say a few words. He began by stating that he was certain that after hearing the presentations, these officials and administrators could not possibly help believing that his firm was the more qualified of the two finalists.

Realizing that all eyes were now upon him, he continued, and gradually gaining momentum, like a football coach at halftime, he spoke louder and louder as he went along. He saw the audience beginning to squirm at this pep talk. Some of them even stared at him in disbelief, and he knew that he had to somehow extricate himself from the situation. But the more he talked, the more panic-stricken he became. He continued, now ranting and raving that no other *$#^&*%$ company could even be compared to his. Minutes later, he lost any remaining sense of propriety and went on using the most vulgar and inappropriate expressions imaginable, until finally he ran out of steam.

Some members of the audience were stunned, their mouths hanging open in disbelief. Others were too embarrassed to look at each other. One high-ranking official, the senior member of the group, simply picked up his brief-case and walked out without a word. The contract went to Firm B, whose proposal was technically inferior and whose bid was slightly higher than that of Firm A.

You will notice that violations of some of the principles discussed below are evident in the foregoing illustration:

Be Prepared

Firm A's loss of this contract was the result of one man's failure to know what he was going to say until he heard the words come out of his own mouth. Unless you are an expert, never try to "wing it" in front of an audience. Know what you are going to say, and know how you are going to say it. Plan the talk carefully, and make sure you have a beginning, a middle, and an end.

Be Confident and Avoid Affectation

This principle could be stated in many ways. It could be said, for example, that you should stand up straight; keep your hands out of your pockets; avoid chewing gum; avoid profanity; beware of catch words, such as "you know," "basically," or "okay"; maintain a dignified bearing; etc. The truth is that all violations of these principles are the result of the speaker's lack of confidence.

Affectation is the speaker's worst enemy, and the too-casual demeanor is the worst of all possible affectations. When speakers use profanity or lean on the podium, affecting a casual manner, they do so in desperation. These actions and word choices are really saying: "Please accept me." When a speaker carelessly chews gum and slurs, "Okay" between every sentence, the real message is, "Can't you see I'm terrified?"

Your body is your most important visual aid. Careful planning and preparation are important, but they will be canceled out if your words and four-color overheads are saying one thing while your body language is saying something else. Audiences want to meet the speaker as well as encounter the subject matter, and these two audience expectations perfectly complement one another. When you move naturally in front of an audience, or even walk about among them, your body language is speaking of friendliness and mutual interest in the subject matter. When you proceed with monotony and stiffness from one overhead to another, you cheat the audience by avoiding this kind of personal engagement.

Use Only Standard English

Slang and vogue expressions create the *illusion* of togetherness, especially among young people. The use of these words and expressions bolsters the confidence of those persons whose identity is tied in with a particular group, whether the group is large or small. It is important to realize that the confidence emanating from such an identity is just as false as the identity itself.

In other words, what works in the group will not work in the world at large. Firm A's vice president later told his co-workers that panic accounted for his fateful speech. He explained that the incident related to something that had happened when he was a junior high school student. He was transferred to a new school where he had felt that same kind of panic every day, and his way of coping was to affect a tough demeanor, thereby hiding his fear and gaining the acceptance of his young peers. He added that at the proposal meeting he had found himself involuntarily using a tactic that had worked in the past.

Respect the Audience

Don't assume that the audience is indifferent to what you have to say. You owe it to them to assume that they are intelligent, inquisitive, and perhaps most important of all, sympathetic. They don't want to see you fail. They want you to succeed, and if you show an interest in your topic, they will do likewise.

Never "talk down" to an audience. When you do so, you not only insult them, but at the same time you call attention to yourself rather than to the topic.

In general, any kind of falseness will be apparent to at least some members of the audience. Therefore, respect for an audience requires that you be yourself. If you are young and inexperienced, being yourself may not be easy. But public speaking can be a maturing experience that will help you to become comfortable with who you really are.

Adjust Your Voice Level To Suit the Environment

Things to be considered are the size of the room, the amount of sound-absorbing material in the room (such as rugs, heavy furniture, drapes), and the level of ambient noise in the room. Look for cues from the audience, especially those at the back of the room. If you don't know whether people at the back can hear you, ask them.

Never try to talk above the noise made by members of the audience who are talking to each other. Stop and look at them. If they don't notice that you have paused because of them, chances are other members of the audience will very quickly let them know they are being rude.

Put Variety in Your Voice

The best way to put variety in your voice is to think about your topic and generate as much enthusiasm for it as you can. If you are interested in the topic, your voice will reflect that interest. Variety will be a by-product that you won't even have to think about. Your interest in the topic will have an important side effect: the audience will catch your enthusiasm. In any event, no audience can be very attentive if speakers are indifferent toward the subject matter of their own talks.

Establish a Smooth Pace

Most amateur speakers talk too fast, and a fast pace can lead to poor enunciation. A normal speaking rate will permit you to utter about 650 words during a five-minute speech. If you begin your speech by talking rapidly, you will find that slowing down is difficult. If, on the other hand, you begin with exaggerated slowness, enunciating all the words clearly, you will be able to increase the rate easily without sacrificing proper enunciation. When you have established a reasonable rate, stick to it.

Also, an uneven pace — one that includes long gaps and intermittent bursts of speed — can become a distraction to the audience. Rehearsing your speech several times before doing it in front of the audience will familiarize you with your note cards and thus help you to keep an even pace.

Maintain Eye Contact with the Audience

Have you ever tried to carry on a conversation with somebody who wouldn't look at you? That avoidance of eye contact is a major distraction. Most members of an audience will take a mental vacation and tune you out altogether if you don't look at them.

Try not to look at just one person, but also be careful to avoid a regular scanning cycle. With practice, good eye-contact technique will become quite natural. Of course, you will have to refer to your note cards, change visuals, and attend to the projector; but the audience will expect these temporary lapses in eye contact and will not be distracted by them. If you maintain eye contact 75 percent of the time and keep individual absences of eye contact from exceeding 10 seconds, you will have done a good job.

Summary

Recalling these principles of good public speaking will help you to make a beginning as a speaker. Perhaps you have some experience in public speaking. If so, you should be able to see how these principles fit in with what you already know.

Industrial Speaking

Some books on speaking cite differences between *impromptu, extemporaneous,* and *"read"* speeches. They cite the impromptu speech as one made on the spur of the moment, the extemporaneous speech as one prepared in advance but made to look spontaneous, and the read speech as one given word-for-word from a script. To avoid the confusion caused by the common use of *impromptu* and *extemporaneous* as synonyms, the term *industrial speech* (instead of extemporaneous speech) is used herein to refer to that class of oral presentations normally given in response to a wide variety of occasions within the working world of the technical employee.

The Impromptu Speech

Most technical employees are normally not called upon to make impromptu speeches as part of the job. On the contrary, most industrial speaking situations are special affairs that require careful planning. Those who do find themselves in a position of having to make a speech on very short notice, however, will have been called upon because they know the most about given subjects.

For example, you may find yourself in a departmental meeting during which the topic of your particular project comes up. If the moderator mentions your name as the one knowing the details of the project's status and calls upon you to bring everyone up to date, you will have five seconds, at most, to gather your thoughts and begin your presentation.

The trick in doing a good job of impromptu speaking is to use those five seconds to decide on an organization for your talk. With your topic in mind, you could decide on an organization by mentally scanning the list that follows:

1. Simple Enumeration
2. Past — Present — Future
3. Problem — Possible Solutions — Analysis — Choice

SIMPLE ENUMERATION

Often, you can base an impromptu talk on this simplest of all possible organizational patterns. You might begin by saying, "I would like to offer four reasons that the project ought to continue on its present course ..." or "Three applications have been identified for the product ...." Once the people in your audience know what structure to expect, they will be able to assimilate what you say quite easily.

PAST — PRESENT — FUTURE

This organizational pattern will work well for explaining a project's status.

- PAST: "We have completed all of the design work ... ELABORATION,"
- PRESENT: "Currently, we are conducting tests on the ... ELABORATION,"
- FUTURE: "The next phase of the project will see ... ELABORATION."

PROBLEM — POSSIBLE SOLUTIONS — ANALYSIS — CHOICE

This pattern of organization permits a complete analysis leading to the best choice from among several possible solutions to a problem. It is an extremely useful pattern for explaining the rationale behind a course of action or, perhaps, for justifying a request for more time, further funding, etc.

The Industrial (Extemporaneous) Speech

Unlike the impromptu speech, which you are called on to make on short notice, the industrial speech is one that you have very carefully prepared. However, the industrial speech is similar to the impromptu to the extent that it is not memorized.

Attempting to recite a speech word-for-word from memory is a game of chance. If you forget where you are, you may have to back up and start again from the beginning. Even if you can get all the way through the speech without missing a word, you will likely not do a good job of communicating with the audience. More often than not, your speech will sound like a "canned" sales presentation, such as the ones memorized by people selling encyclopedias and vacuum cleaners door-to-door.

Remembering and communicating are two entirely different activities. Therefore, prompting your memory with carefully prepared note cards, thus removing the burden of recalling a memorized text, will allow you to concentrate on communicating with the audience. And communicating with the audience is the purpose of every speech.

The "Read" Speech

The "read" speech is one that you read from a script. The odds of successfully communicating with an audience by means of a read speech are even longer than by means of a memorized speech. For this reason, reading a speech for any work-related purpose is considered inappropriate in most organizations.

The Differences between Industrial Speeches and Those Given for Other Reasons

Industrial speeches differ in several important ways from classroom lectures, after-dinner speeches, valedictory addresses, and so on.

Classroom lecturers do not usually have to worry about a time limit. If someone asks a good question, the instructor may choose to spend the whole class discussing its ramifications, picking up the thread of today's lecture tomorrow or the next day. Industrial speakers do not have this luxury. If they try to handle questions in the middle of the speech, they can easily lose control and never finish saying what they have to say. Classroom lectures also allow for more familiarity with the audience than do industrial speeches.

The form of the industrial speech also differs markedly from the after-dinner speech, which is often begun by addressing the audience and telling a funny story. Addressing the audience is inappropriate in most industrial speech settings. And leave out the humor unless you are an accomplished storyteller. If your story falls flat, so will your speech.

Valedictory speakers also begin by addressing the audience: "Members of the faculty, honored guests …," and so on. Industrial speakers, on the other hand, begin with a self-introduction (provided that the speaker has not already been introduced by a moderator), a bit of background information, and an overview of the organizational pattern that the speech will follow.

Preparing a Speech

Two important facts must be considered when preparing a speech:

- Members of an audience remember what they see longer than what they hear.

- Members of an audience remember details by reference to an overall structure.

Heeding the following suggestions will help you make good industrial presentations. These suggestions are intended as specific guidelines developed from the principles of good public speaking mentioned earlier.

Visuals

Studies have shown that an audience remembers what it sees much better than what it hears. Accordingly, good prepared speeches often include visual materials, and you should always consider using visuals when preparing a speech. The first thing you must do is find out what kind of equipment is available at the site of your speech. Overhead projectors are used widely, as are slide projectors and flip charts.

Flip charts are useful for speaking to small groups in small rooms, but do not afford an image area large enough for presentations to large groups. In addition, they are not easy to handle well.

Slide projectors offer excellent reproduction quality, but require a darkened room, a disadvantage that is often worth tolerating if the visuals are necessary or if such equipment is all that is available.

The overhead projector is a widely used piece of equipment that affords the speaker many options. Photographs and line drawings can be easily reproduced on transparencies by means of several different inexpensive processes. And transparencies can even be made by hand, using a variety of colored pens, tapes, and papers. Thus, the speaker can easily place key points of the speech on a transparency to keep the audience oriented at all times.

Some speakers even commit to transparencies all of the points to be made, thereby eliminating the need for note cards. One must be careful, however, to avoid putting too much on one transparency. A good rule of thumb is to place no more than six lines of type on each, making certain that all written material is in letters no smaller than 18-point (equal to one-quarter inch or 6.35 millimeters).

Two more advantages of the overhead projector are its strong light, which permits its use in a fully lighted room, and its orientation in the room, which allows the speaker to remain facing the audience at all times.

Plan to use visual aids if equipment for showing them is available. Visuals can be used in exactly the same way they are used in technical writing situations. (See Chapter 6.)

Structure

BEGINNINGS

The beginning of a speech must orient audience members and prepare them for what they are about to hear. Included in this orientation are a clear statement of the subject matter and a well-defined description of the structure of the speech. A typical opening follows:

Good morning. I'm Sharon Wilder of the Engineering Department, and I've been asked to give you a briefing on our plans for designing the Model B Simulator. We see the best course of action to be composed of a three-stage development, the first stage being a VFR prototype, followed by the introduction of motion simulation and full avionics in the second and third prototypes, respectively.

Our efforts in developing the VFR unit will be concentrated in computer-generated ...

In the foregoing example, Sharon Wilder has managed to get directly into her topic with a minimum of preamble. In very few words, she has greeted the audience, identified herself, stated the nature of the topic, and described the structure of the talk. (The talk is going to be in three parts: plans for each of three prototypes.)

It should be noted that she did not begin by saying, "Ms. President, Mr. Vice President, and members of the Executive Council." Addressing an audience in this fashion is inappropriate in most working environments. She settled for a simple and perfectly appropriate greeting: "Good morning."

She identified herself, but she would not have done so had she been known by everyone in the room or if she had been introduced by another speaker. The next part of that same sentence establishes the topic she is going to talk about.

Her next sentence gives a brief overview of the structure of the talk, after which she is able to launch herself directly into the first part of it. The audience is aware of the three-part structure and will be better able to recall details later because they will have a framework to relate them to.

MIDDLES

The middle of the speech must meet the expectations raised in the overview. If Sharon Wilder were to talk about quality control methods in the middle of the speech begun in the foregoing example, she would throw her listeners off track by including a fourth part in a three-part speech.

The middle should be carefully structured to reflect the broad overview given in the beginning, and it should provide substructural information as necessary. Coherence, transitions, and all the other structural elements necessary for a well-prepared written communication are of utmost importance in a speech as well. In fact, these elements are even more important in a speech than they are in a written communication. Readers can always re-read what they have been unable to understand the first time, but listeners, perhaps believing they have missed something, are often reluctant to ask the questions raised by poor structure. If the structure is very poor, listeners will often not even know what questions to ask.

ENDINGS

A good ending concludes, reiterates, explains, and sometimes recommends. It is not only a graceful way for the speaker to get off the platform, but a chance to bring home the significance of what was said.

Finally, most speeches occasioned by work end with an invitation to the audience to ask questions. It is at this point that speakers can easily do a poor job. Some speakers, who have not listened carefully enough to a question, ramble as though talking to themselves or, worse, give the entire speech again in slightly amended form. It is most important to understand each question and to answer only that question.

The question period can become awkward in yet another way. The speaker may wrongly assume that there will be questions and wait too long for a question to be asked. If you have done a good job of communication, there may be no questions. Accept that possibility and, having invited questions, do not wait more than five seconds for a question to be asked. If no one asks a question, thank the audience and leave the platform. Otherwise, you can find yourself a pathetic figure, standing and waiting for someone to ask you a question. Of course, someone will finally take pity on you and ask something, but at this point the whole character of the speech and speaker will have changed in the eyes of the audience. You will have, in effect, relinquished control of the speech, which will subsequently die a slow and agonizing death.

Remain in control to the end of the question period by answering each question precisely and directly. If you don't know the answer, say so and offer to supply the information at a later time. If the question lies outside the boundaries of the topic as you have stated it, say so.

Endings that include a slide presentation during which the lights must be turned off require that the question period occur at the conclusion of the slide presentation. If questions are invited beforehand, questions that may be generated by the slide show will go unanswered, or a second question period will be required. Neither of these eventualities is desirable.

Exercises

14-1. Speech Assignment

The time limit can be specified by your instructor. Try to come as close to the specified time as you can without going over it.

1. Choose a topic that you believe will be of interest to the class. It should be related to your field of study. (If your instructor concurs, you may, as an alternative to choosing a topic, select an article from a periodical, and then proceed with Step 2.)

2. Prepare a brief written outline of the major divisions of the topic.

3. Prepare an introduction, including your name, the title of the speech, the objective of the speech, and an overview of the topic.

4. (Optional) Following your outline, write out the entire speech if you think that seeing the whole thing on paper will help you to get more familiar with the topic. However, don't bring this copy with you when you deliver the speech.

5. Prepare a conclusion by reiterating main points or indicating the significance of what you have said.

6. When you have completed the foregoing tasks, prepare note cards (3" by 5" work very well) that give the major points of the speech. Do not write out the entire speech on the cards. Remember that an industrial speech is not read to the audience. To be effective, it must have an element of spontaneity and be a true communication between the speaker and the audience.

14-2. Using a Speech Checklist

Before delivering the speech you prepared in Exercise 14-1, answer all the following questions and make the necessary changes.

1. Do I have a beginning, middle, and end?

2. Have I included enough background information in my introductory statement to adequately lead my audience into what is to follow?

3. Does my overview give a good framework so that the audience will be able to understand my presentation?

4. Does the middle of my presentation accurately follow my overview?

5. Does my conclusion follow logically from the material I have presented?

6. How long did I take to give my presentation during practice sessions? How does this time compare with my assigned time allotment? Do I need to add or subtract material?

7. Are all my note cards and visuals prepared?

8. Do I have a sufficient command of my material necessary for confidence?

9. Am I ready for any pertinent question that may arise?

10. How do I look? Am I comfortable with my appearance and clothing?

CHAPTER 15

Job Search Techniques

OBJECTIVES: In this chapter, you will learn about résumés, letters of application, and interviews

The Résumé

The résumé is a document that tells a potential employer of your qualifications. An important point to remember about résumés is that they often tell more than their writers intend to say. Since the résumé represents you, the people who read it will equate its character with your character. If it is overstated, potential employers will think you are boastful and overbearing. If it contains misspelled words and grammatical errors, they will think you are careless or illiterate.

For these reasons, you should put your résumé together with utmost care. And in order to do a good job of writing a résumé, you will also have to know a few things about graphic design.

The Graphics of a Good Résumé

Glance quickly at this page, without actually reading it. Did you notice that your eyes moved from the upper left-hand corner of the page to the headings and then to the underscored lines? These lines make up the graphic "surface" of the page. Important information is conveyed by this arrangement. In other words, the reader doesn't need to read the entire page to get an idea of what it contains. Now consider the résumé shown in Sample 15-1. Don't read it yet; just glance at it.

SAMPLE 15-1	Résumé Exhibiting Poor Graphic Design

Shirley J. Newsome
1245 Marcy Place
Danby ON N9J 4V3
Telephone: 519 555-9876

Education:
1990–1992. River Valley College, Danby, Ontario. Diploma in Architectural Technology. Second-year Option: Interior Design.

continued →

Experience:

<u>1997– present</u>. Sally's Interior Design, Ltd., Danby, Ontario. Interior Designer. Analyze customer needs and requests for color coordination and furniture selection and arrangements in professional offices.

<u>1992–1997</u>. Acme Designs for Living, Rimbey, Ontario. Interior Layout Technician. Duties included planning of interior layouts of custom-built houses. Supervised two apprentice designers from 1995 to 1996.

Interests and Activities:

Violinist in amateur music groups, including the Danby Youth Chamber Ensemble.

What can you remember about the example shown in Sample 15-1? Probably not much. From a graphics standpoint, the most prominently featured items are dates. The dates have some significance, but they are certainly not of primary importance. A résumé can be arranged in chronological order without emphasizing the dates, as shown in Sample 15-2.

SAMPLE 15-2 | **Résumé Exhibiting Good Graphic Design**

SHIRLEY J. NEWSOME
1245 Marcy Place
Danby ON N9J 4V3
Telephone: 519 555-9876

EDUCATION

RIVER VALLEY COLLEGE, Danby, Ontario. 1990–1992.

<u>Diploma in Architectural Technology</u>, Second-year Option: <u>Interior Design</u>.

EXPERIENCE

SALLY'S INTERIOR DESIGN, LTD., Danby, Ontario, 1997– present.

<u>Interior Designer</u>. Analyze customer needs and requests for color coordination and furniture selection and arrangements in professional offices.

continued ➔

ACME DESIGNS FOR LIVING, 1992–1997, Rimbey, Ontario.

<u>Interior Layout Technician</u>. Duties included planning of interior layouts of custom-built houses. Supervised two apprentice designers from 1995 to 1996.

INTERESTS AND ACTIVITIES

Violinist in amateur music groups, including the Danby Youth Chamber Ensemble.

No new information is included in Sample 15-2. The personal information group is centered, and the name of the person is in all-capital, boldface letters. Capital letters are also used to indicate the names of employers, and boldface type is used to show the overall framework. A liberal use of "white space" completes the bag of graphic tricks. Combined, these visual characteristics make for a document that conveys information to the reader on two levels: a "surface" level that does not require meticulous attention, and a second, deeper level that supplies other relevant information.

Another kind of résumé, called the functional résumé, enjoys some popularity. Such a résumé groups all the information in job function categories, such as supervision of personnel, field service work, working with the public, or some other general area. Company names and dates are usually omitted. The advantages of the functional résumé are few, and it should be noted that many employers regard it as an exercise in camouflage. A carefully structured graphic presentation of a chronological résumé is probably the best choice for most people.

The Contents of a Good Résumé

The purpose of the résumé is to interest the employer enough to invite you to a personal interview. That purpose must be kept in mind when you are deciding what material is to go into it. Also, you should understand that the résumé is not an application form. Most firms require prospective employees to complete application forms, which are quasi-legal documents containing a statement certifying the truth of the information given. These forms are quite specific as to the information required; they also usually direct the applicant to account for all times of unemployment.

Although the résumé must not contradict the application form, which will be required before employment with the firm begins, the writer does have some latitude in selecting the material to be included in the résumé. Periods of unemployment need not be (should not be) accounted for in a résumé, and all jobs need not be included. Likewise, although an application form will require the name of the high school attended, it is usually undesirable to include such information in the résumé unless you have no post-secondary education to mention.

Ask yourself the following two questions when considering an item for the résumé:

- Will this item help the employer to judge my suitability for the position?
- Will this item make the employer want to talk to me in person?

If the answer to both questions is "yes," include the item. If the answer to one question is "yes," you should *probably* include the item. If the answer to both questions is "no," omit the item.

Arrange the major headings in the résumé in the order of their importance. That is, if you have just finished your schooling, place the education heading immediately after the personal group at the top of the page. Likewise, information under headings should be arranged in a most-recent-first, next-most-recent-next chronology.

Each heading category is discussed in the following paragraphs.

PERSONAL INFORMATION

The first information group should always include your name, address, telephone number, and fax and e-mail numbers if you have them. Some résumés include an additional "personal data" category that gives particulars of marital and citizenship status, physical characteristics, and so forth. Include this category if you believe it will help the employer to judge your suitability for the position. For example, if Canadian citizenship or landed immigrant status is a prerequisite for a position, as it is for many federal and provincial government jobs, you should include a statement affirming your status, at least when applying for such positions. Or if you are applying for a job in a remote site in the Yukon, family status might be an important consideration.

EDUCATION

All post-secondary education directly applicable to the position applied for should be listed, including special courses taken outside of a regular program. Also, include under this heading any option courses, prizes, or scholarships.

Post-secondary degrees and diplomas, even though not directly related, should usually also be included, since these can provide a good insight into the overall interests and past pursuits of the applicant. Extension courses taken purely for interest's sake may be included or not, depending on the courses themselves and the image you are trying to convey.

A mention of secondary education doesn't very often add anything but a distraction to a résumé, although attendance at a particularly prestigious high school may be worth mentioning in some cases. Of course, you will want to mention secondary education if you have nothing else to put under the education heading.

A list of courses and grades is inappropriate unless the employer to which you are sending your résumé asks for such a list. Applicants for employment must always be responsive to the advertisement or bulletin citing the position.

EXPERIENCE

This category, which is sometimes also called *Work Experience* or *Positions Held,* should begin with the present or most recent job and progress back through to the first job. All jobs need not be mentioned. As shown in Samples 15-1 and 15-2, you should give a position title and a brief description of your duties and responsibilities. Make the description of duties as brief and as specific as possible. Avoid the expression *involved in,* since it could mean anything. Action verbs, such as *planned, operated, designed, built, coordinated,* and *supervised,* are good foundations on which to build descriptions of duties.

Exercise special care to assure that parallel structure is maintained within the job descriptions. Don't write, "Designed transmitter circuits for electronic locator transmitters ..." in one description and, "Marker beacons ..." in another. You would force the reader to wonder what you had done when you were working with marker beacons. Did you design them? Did you test them? Did you service them in the field?

INTERESTS AND ACTIVITIES

Every résumé writer must be careful in choosing items for this category. Many people are inclined to list every activity they have ever tried. A person who lists too many items under this heading creates a picture of someone who won't have time for employment. Be especially diligent in omitting such entries as "listening to my stereo," an activity that certainly isn't going to impress many prospective employers.

List activities that really are activities. For example, don't list "music" unless you play a musical instrument. Everybody listens to music. Don't list football unless you are an active player. In any case, be specific about the activity. Say, "Play linebacker for the Calgary Buffaloes Junior Football Team," not just, "Football."

Remember that the purpose of this category is to show that you are a well-rounded person active in community affairs. If you don't have any interests outside of work, omit the category.

OTHER CATEGORIES

The professional organizations you belong to may be listed under a separate heading. A list of organizations to which a person belongs helps to show professional interest that goes beyond the confines of the workplace. A heading called *Publications* is another category that should be included by those persons who have published papers, articles, or books.

A *Recognitions and Awards* heading can sometimes be included separately also, although most such achievements would be associated with one of the other categories, such as education, and would be better mentioned there.

A *References* category is often included in résumés. However, most employers pay no attention to references listed in résumés and would prefer not to see them included. The reasons are simple enough. A large firm may receive several hundred résumés in response to an advertisement for one position. Writing or telephoning the references in order to decide whom to interview would be impossible.

However, you will have to include references on the application form, and your prospective employer will very probably question the people you have listed after placing you on a short list. You should, therefore, know what references you plan to use and get permission from each of them in advance. If you wish to do so, you may indicate on your résumé that references will be supplied on request, although most employers will regard such information as superfluous.

References are usually checked as a final step in the hiring process. If any are negative, the employer will go on to the next candidate in line. In other words, poor references may be instrumental in preventing your getting the job. Good references are essential.

Some résumés include a *Job Objective* heading, usually as a first category. The trouble with this category is that it can often mislead. Is this the applicant's present job objective or ultimate job objective? It is usually better to omit this category when applying for a specific job in response to a specific request. Let the employer know in the cover letter accompanying the résumé what advertisement you are responding to.

A Student Résumé

The problem faced by most students is that of having little or no relevant experience to include on a résumé. Employers realize that students have to begin somewhere, since nobody is born with experience. Don't be ashamed of the jobs you had to do to get money for your education. These jobs show your sense of purpose and responsibility, and you should include them with pride. If you bagged groceries or grilled hamburgers, say so within a brief description of duties. If you are fortunate enough to have gained some experience related to the field in which you are applying for work, include as much detail as you think will be helpful.

Sample 15-3 shows a typical student résumé.

SAMPLE 15-3 | **Student Résumé**

SHIRLEY J. NEWSOME
1245 Marcy Place
Danby ON N9J 4V3
Telephone: 519 555-9876

EDUCATION

RIVER VALLEY COLLEGE, Danby, Ontario, 1990–1992.
- Diploma in Architectural Technology, 1992.
  Second-year Option: Interior Design.

EXPERIENCE

SALLY'S INTERIOR DESIGN, LTD., Danby, Ontario, 1997– present.
- **Design Trainee**. Assisted designers in assessing requests for color coordination and furniture selection and arrangements in professional offices.

A&W RESTAURANT, Brandon, Manitoba, 1992–1997.
- **Waitress**. Waited on tables and acted as cashier one day a week.

CARIBOU PARKS AND RECREATION DEPARTMENT, Dawson, British Columbia, Summers of 1990 and 1991.
- **Umpire**. Umpired Little League games.

continued →

The Letter of Application

Form

The letter that conveys your résumé must conform to good business letter format (see Chapter 9). You must also take great care to respond in the manner specified by the employer. When answering an advertisement or employment bulletin, read carefully the part that gives the name and address of the person to whom letters of application and résumés are to be sent. Check carefully to see if you are asked to cite a competition number or similar designation. If so, a subject line may be a good place to comply with such a request.

Occasionally, an employer will ask that the letter or résumé take some special form. A few may require a response in your own handwriting. However, never submit handwritten material unless specifically requested to do so. Follow all of the directions as closely as you can.

Content

The letter of application is a personal communication that accompanies the résumé. It does three things:

1. It introduces the résumé, citing the source of your knowledge of the position (e.g., advertisement, employment agency, friend, etc.).

2. It highlights a part of the résumé that especially signifies your qualifications, and it mentions any significant fact not included in the résumé.

3. It declares your availability and willingness to attend an interview.

INTRODUCTION OF THE RÉSUMÉ

This part of the letter can be a single statement, although two statements are sometimes required to avoid awkwardness. You should be specific about the source. Merely stating that you are responding to a newspaper advertisement for a certain job may give the employer enough information to correctly channel your application, but giving a complete citation will help in another way. That is, company recruiters spend large sums of money in advertising open positions, and many of them measure the number and quality of responses against the amounts spent, in order to compare publications. Giving a complete citation will show the employer that you are aware of such matters.

HIGHLIGHTING THE RÉSUMÉ

Only you know what you should emphasize in this paragraph. Perhaps a mention of a past job experience that you thought was particularly useful to you in choosing your career would be appropriate. Or perhaps a statement concerning your reasons for choosing a particular course option would be helpful. In any case, the statement should be concise and purposeful. Also to be included in this paragraph would be the mention of any significant fact not included in the résumé. Usually, however, an up-to-date résumé will cover all important information.

DECLARING YOUR AVAILABILITY FOR AN INTERVIEW

The whole purpose of sending a letter of application and résumé is to get an invitation to an interview. Reason therefore demands that you end the letter with a mention of the interview. You may simply say, "May I come in for an interview to further discuss my qualifications?" The more subtle approach is to state your availability for an interview at a time convenient to the employer.

There is no need to end the letter by telling the employer how to get in touch with you. Your name, address, and telephone number are present in the personal information group on the résumé. Repeating them in the final paragraph of the letter sounds a note of anxiety on your part. Also to be avoided are self-conscious statements, such as, "I am anxiously awaiting your reply." The employer knows that a prompt response is important to you. If an employer doesn't know (or doesn't care), a display of anxiety will not help.

ADDITIONAL ATTACHMENTS

Students often ask whether to attach copies of diplomas, letters of reference, and like materials in addition to the résumé. A survey of employers indicated that very few believe this practice to be a good idea. A personnel manager for one large firm stated that she immediately detaches all such material and throws it into the wastebasket.

Sample 15-4 shows a letter of application that might accompany the résumé shown in Sample 15-3.

SAMPLE 15-4 Letter of Application

321 Parliament Street
Regina SK R7U 2K3
August 7, 1998

Ms. Rita Palmer
Employment Coordinator
Acme Designs for Living, Ltd.
3040 Water Street
St. John's NF A9Y 2U4

Dear Ms. Palmer:

The enclosed résumé is submitted in response to your advertisement in the August 6 edition of the *St. John's Gazette*. The advertisement stated that two positions were available for interior designers.

Following my graduation from River Valley College, where I was awarded the Diploma in Architectural Technology (Interior Design Option), I spent one year in training with Sally's Interior Design in Danby, Ontario. My hope is to continue my career in an environment that offers new challenges and opportunities for learning. I would be pleased to meet with you at any time in your St. John's offices to further discuss your requirements and my qualifications. Thank you.

Yours truly,

Shirley J. Newsome

Shirley J. Newsome (Ms.)

Enclosure: Résumé

The Interview

If your application is looked upon with favor, you will be invited for an interview with one or more representatives of the company. This stage of the employment search process presents special challenges. Failing to prepare for an interview is the quickest way to eliminate yourself from any further consideration. The following list of suggestions should be helpful:

- *Be certain that you know when and where to appear*. This suggestion is particularly important if you must travel to another city for the interview. You could find yourself in the wrong place at the right time. Plan to be early in case you require extra travel time.

- *Be well groomed*. Personal appearance counts for a great deal to the average interviewer, who will notice how you look before asking you any questions. It is impossible to make up for an unkempt look. You could be the most skilled and meticulous worker in the world, but if your appearance doesn't reflect those qualities, few prospective employers will believe you have them.

- *Be well informed*. Find out as much as you can about the company. What goods does it produce, or what service does it perform? How is the company organized? What are its major markets? Who are the competitors? What is the interviewer's name, and how is it spelled and pronounced? Many interviewers conclude the interview by inviting the applicant to ask questions. It is a good idea to prepare a few intelligent questions for this contingency. Also, know what you want. At what salary do you expect to begin? What are your long-term goals?

  Questions about salary expectations can create a special difficulty. How much should you ask for? You don't want to price yourself out of the market, but neither do you want to accept less than the company would be willing to pay. The best answer is one that reflects the current salary schedule of the company. Unfortunately, this information is not always readily available, in which case at least two other responses are possible. You could cite the lowest salary you would accept based on the desirability of the job; or you could answer the question with your own question, "How much does the company normally pay an employee with my qualifications?"

- *Be well equipped*. Take along copies of diplomas, certificates, samples of work, letters of reference, etc. The interview is the proper time to bring out the heavy artillery.

- *Be well rested.* An interview can sometimes be an arduous experience. Some firms convene panels of people to conduct interviews, and such interviews require the interviewee to juggle several lines of questioning at once. An interview can last anywhere from several minutes to several days. As a routine practice, at least one major company conducts interviews that last for eight full hours. Some companies also administer a battery of aptitude and psychological tests either before or after the interview.

- *Be aware of the possibility of encountering unskilled interviewers.* Unfortunately, the unskilled interviewer is not as rare as one would hope. Some interviewers think that the best way to judge prospective employees is by treating them with rudeness, indifference, or cross-examination techniques. These practices, although they reveal more about the shortcomings of the interviewer than anything else, can be unnerving. Remain unemotional and confident if you should encounter such tactics.

  Those unskilled interviewers who are not rude may be uncomfortable and not know what questions to ask. You can often take control in such an interview by trying to put the interviewer at ease and by volunteering as much information as you can. Make the interviewer's job easy, especially the interviewer who regards the process as distasteful. In so doing, you will create a positive memory of the encounter.

  The skilled interviewer will put you at ease by regarding you as an equal and will display a friendly and helpful attitude.

- *Be yourself.* Be friendly, but be aware that a too casual attitude is the worst kind of affectation. An applicant's stuffy or too formal approach may mildly amuse an interviewer, who might be understanding enough to overlook such a lack of directness. Applicants who are, on the other hand, flippant or familiar in their manner exhibit unacceptable behavior.

- *Be confident.* You have something to offer, and the employer has something to offer. You are getting together in order to find out whether an association will be a good thing for both of you. Do not, however, be overconfident. A prospective employer may quite justifiably believe that the training you have had in school is merely a foundation on which to build. If you think you "know it all" already, you will not make a good impression.

- *Be alert.* Listen to each question carefully, and try to answer it fully. Avoid the one-word answer most of the time. Interviewers appreciate cooperation in their attempts to elicit the information needed to judge an application.

  Be alert also to the employees you encounter outside of the formal interview. Sometimes these persons are consulted concerning the attitude you display when your "guard is down." One employer was known to have a personal secretary who would engage applicants in informal conversation during the time they were waiting to go into the interview. This person would then report her findings to her supervisor, who placed great weight upon her assessment.

- *Be prepared for an "It's-your-move" opening.* Some interviewers will begin with questions such as: "So, you want to work here, do you? Why?" "Tell me about yourself." "So, What's your story?" (Refer to the points made above about unskilled interviewers.) You should be ready to answer such a question by having prepared a brief speech, citing your qualifications as they appear in the résumé and briefly describing your goals and expectations.

Exercises

15-1. Writing a Résumé

1. Write your own résumé in response to an advertisement for a job for which you are now qualified.

2. Write your résumé in response to a job you think you might qualify for five years from now. Make up the experience section, but be reasonable.

15-2. Writing a Letter of Application

Write a letter of application to accompany one of the résumés you wrote in Exercise 15-1.

15-3. Answering Interview Questions

Write out answers to the following questions. Elaborate as much as possible on each question, giving reasons for your answers when the question implies that you should do so. Some of the questions are unfair and somewhat devious. Try to determine the intent of the questioner, who might be one of the "unskilled" interviewers mentioned above.

1. What kinds of work do you like best?
2. What was your favorite subject in school?
3. What subject did you dislike most?
4. What do you believe to be your greatest strengths?
5. What do you believe to be your greatest weaknesses?
6. What do you expect to be doing five years from now?
7. How important is salary?

8. Do you mind taking orders?
9. What kind of people annoy you?
10. Have you ever had personal conflicts with bosses or fellow workers?
11. Did any of your former employers treat you unfairly?
12. Who influenced you most during your school years?
13. Why did you choose _____ as your field of study?
14. What are your ultimate goals?
15. What aspect of your present or last job did you find most distasteful?
16. What job did you like best?
17. Do you live with your parents?
18. How much did you contribute to the cost of your own education?
19. You have listed _____ on your résumé as an interest; what can you possibly see in such a hobby?
20. Do you like to work with other people, or do you consider yourself to be a loner?
21. This company is lean and tough, and we don't have time for the employee who can't keep up. How do you think you would fit in here?
22. This company likes to bring its employees along slowly. You won't get a promotion here until we know you're ready for it. Does this conflict with your timetable for personal advancement?
23. I'll be honest with you. We have someone already in mind for this job. Do you want to go through with this interview anyway?
24. Tell me about yourself.
25. Do you think that the grades you were given in school accurately reflect your abilities?
26. What books have you read lately?
27. What periodicals do you subscribe to or read regularly?
28. Tell me in 25 words or less how you feel about the quality of the education you have received.
29. The people who make it with this company work long hours and always put the job first. Would you have a problem with that?
30. I always say it's who you know and not what you know. What do you think?

15-4. Making Up Interview Questions

Write out as many interview questions as you can, at the same time formulating your own answers.

15-5. *Interview Practice*

Using an advertisement or job bulletin related to your field of study as a basis for interviews, convene a panel of interviewers and collect the letters of application and résumés of all of the other students in the class. Have the panel select the best five and interview them, having carefully prepared the questions and strategies in advance. Choose freely from the questions listed in Exercise 15-3 and the questions generated in Exercise 15-4. Feel free, also, to create new questions based on individual résumés.

Ethics

CHAPTER 16

Ethics and the Technical Writer

OBJECTIVES: In this chapter, you will learn about the requirements for a system of ethics. You will also learn that ascertaining the facts of situations with ethical implications is the first step to acting ethically.

You will then learn about the ethical obligations that technical writers have to their employers, their employers' clients and customers, and the human community at large. Working through the exercises at the end of the chapter, you will assess situations and apply the facts discovered in defining the ethical implications of each case. You will also learn to recognize what constitutes the fair use, as opposed to the abuse, of rhetorical principles.

Ethical Foundations

Introduction

Imagine for a moment that you are a technical writer working for a company that has a large research and development contract and that you are in charge of preparing the production copy of an interim report. All the members of the research team have given you draft copies of their findings thus far, and all the findings are negative. The inescapable conclusion is that the project is not feasible and that any further work will be a waste of the customer's money. It is clear that if you do a proper job, the project will be canceled. Your boss, a corporate vice president, instructs you to rewrite the copy to convey the false impression that the work done thus far is inconclusive and that the prospects for eventual success are very high. A dark hint is included with the instruction: if the report is accurate, you and a lot of engineers and technicians are going to be out of work. What do you do?

A. Ignore the boss's instructions, write the truth, and take your chances.

B. Follow the boss's instructions, start looking for another job right away, and resolve to reveal the truth later.

C. Follow the boss's instructions. Who is to say what's true anyway. Maybe the research team is wrong. Anyway, it's not just your job that's at stake. You have an obligation to your colleagues.

If you chose *A*, you are courageous, ethical, and uncompromising. If you chose *B*, your actions are certainly understandable. If you chose *C*, you have compromised yourself in a special way. That is to say, you have rationalized your decision, and when you rationalize, you lose something very important in the process. Some people call that important something *personal integrity*.

Let's throw more light on our example by extending it to a historical situation. Suppose that, instead of a technical writer in present-day Canada, you are a junior officer in Hitler's SS in the 1930s. Suppose that the job you are told to do is to round up all the citizens in a certain ethnic neighborhood, load them onto trucks, and take them to the nearest railway terminal for shipment to an extermination camp.

You are fully aware that choosing the equivalent of *A* could easily get *you* a one-way ticket to the extermination camp. In other words, doing what you know to be right will probably cost you your life. Would you be justified in asking yourself whether such a gesture (these people are going to the extermination camp whether you obey the order or not) is the smart choice? Would you lose your integrity by choosing to obey the order while silently resolving to fight against this evil policy at a later date? Perhaps you would decide to save your-self and take up the fight in a way that would be more practical and effective. In other words, you might reasonably conclude that a simple confrontation you can't possibly win will not benefit the people you would like to save. Therefore, choosing *B* would, as in the less drastic example of the false report, be understandable. A future war crimes tribunal might well take your predica-ment into account, particularly if you had followed through on your silent promise to do something about the grotesque policy of your government.

Choosing *C*, however, makes you a conspirator in the act of genocide. What is the distinction between *B* and *C*? How can there be such a differ-ence? The difference has to do with the foundation of your decision. All eth-ical acts proceed from a knowledge of right and wrong.

Ethics and Epistemology

We need to begin by defining the two terms in our heading. Ethics is the branch of philosophy that deals with how we should behave. Epistemology is the branch that deals with how we come to know things. You probably see the connection already. That is, you can't act in accordance with the values that govern ethical behavior unless you first acknowledge the values themselves. The most important values for the ethicist are truth and justice, and those who wish to act ethically in any given situation must ask the following questions:

- What are the facts in this case?
- What action will serve justice in this case?

The answer to the first question is not always easy since the facts of a case are not necessarily self-evident. Some people are inclined to leap to conclusions and create false dilemmas for themselves. You will learn more about the process of assessing situations as you work through the exercises at the end of this chapter.

Answering the second question demands an accurate definition of *justice*. What is justice, and how can we know what action will serve it in a given case? Is justice served automatically when the law is obeyed? We must give a *no* answer to that question, and many historical examples illustrate why. For instance, the laws of totalitarian states have been known to sanction crimes against humanity, such as genocide, as was the case given in the example above. We may also observe that, apart from totalitarian regimes whose rulers use their legal systems for their own purposes, even democratic countries have unjust laws. Fortunately, most such laws don't lead to crimes against humanity at large, but they nevertheless provide for the unjust treatment of some people.

So, what is justice? Some theorists claim that we human beings invented it so that we could live together and formulate laws. But if that theory were true, we would have no basis for condemning any legally sanctioned atrocity, no matter how horrible we might think it is. If justice were a human invention, in other words, we would have to conclude that Hitler's policy of genocide served justice at that time and in that place.

Clearly, then, justice is not a human invention. But if it isn't a human invention, what can it be? Consider the following example.[1]

> On the first day of classes, an ethics instructor walked into the classroom, eyed each of his students coldly, and said, "I'm going to pass everybody in the three rows nearest the door and fail everybody in the three rows nearest the windows." After a stunned silence, hands shot into the air. The instructor invited one student to speak.
>
> "That's not fair!" the irate student shouted.
>
> "Who says so?" the instructor shouted back. "I'm the instructor in this class, and I'll decide who passes and who fails. And who says I'm being unfair? I resent your statement." At the end of a heated exchange that lasted the whole period, and during which all but a few of the students joined in, the instructor smiled and said, "Congratulations. You have just had a glimpse of justice and have made some pertinent arguments in support of it. I have no doubt that all of you will pass this course with flying colors."

These students had "seen" justice because their instructor had thrown it into sharp relief by declaring his intention to breach it. They did not have to make reference to an artificially constructed concept, and neither did they have to "re-invent" a concept in order to comprehend the fault in their instructor's stated intentions. As their ethics class progressed, most of the students learned that justice is not only real, but real in a way that even physical objects are not.

Ethics and the Law

A FICTIONAL CASE STUDY[2]

> Under federal contract, Pitbull Industries, Inc., designed a power plant for use in a remote northern community. Pitbull subcontracted the Start-Up, Operating, and Maintenance Instructions Manuals to Skylark

---

[1] For a number of years, students in Earth Resources Engineering at the Northern Alberta Institute of Technology chose, as part of the required program, one of several option courses, which included paleontology, history, and philosophy. The author taught the philosophy option and, one year, opened his class with the following "exercise."

[2] Though fictional, this case study was inspired by a real incident.

Publications, Inc. After Pitbull had completed the design and had received full payment for its work, the federal government decided not to go ahead with the project.

Pitbull immediately canceled Skylark's contract for the manuals and refused payment, even though the manuals were ready for delivery. In refusing to pay, Pitbull cited a contract clause that read as follows: "Acceptability of the manuals shall be at the sole discretion of Pitbull Industries, Inc." Skylark argued that the clause obviously meant only that substandard work would be refused, not that Pitbull could cancel the contract to enhance its own profits at the expense of its subcontractor.

Pitbull, a very large company with a permanent legal staff, "stonewalled" Skylark, a small contractor with few financial resources. Relying on Pitbull's promise of payment agreed to elsewhere in the contract, and confident of its own ability to deliver acceptable work on time, Skylark had financed the entire project itself. When Pitbull refused to honor the contract, Skylark was forced into bankruptcy.

Case Analysis

The foregoing case illustrates how a legal system may fail to ensure justice even if the law itself is just. Skylark Publications, Inc., could not afford to pursue a lawsuit against a large corporation such as Pitbull Industries, Inc., even though it would most likely have won eventually. Pitbull's executives knew that Skylark lacked the financial resources to carry on a protracted court battle. On the advice of their legal staff, these executives based their decision to refuse payment on "bottom-line" economics.

So-called "bottom-line" economics and ethical standards cannot coexist. History has also shown that the maximum-profit-above-all-else philosophy is a poor way to conduct business. Compare the Pitbull case study with the one that follows.

A TRUE CASE STUDY[3]

In the 1960s, the Westinghouse Astronuclear Laboratory was developing the NERVA reactor, part of a space engine that would propel heavy payloads into deep space. After the design was essentially complete and prototypes had been successfully ground tested, Westinghouse received notice that the next phases of the project were canceled. Cancellation of the project meant that Westinghouse would have to lay off hundreds of engineers and scientists.

---

[3] The author worked as an engineer/technical writer at the Westinghouse Astronuclear Laboratory in the 1960s.

The Astronuclear Laboratory executives took immediate action. They put all their personnel representatives on overtime to help find jobs for the employees who would have to be released. Personnel representatives queried other Westinghouse divisions and found positions for many whose jobs had been discontinued, and they prepared and mailed out résumés for those remaining. Line managers and executives telephoned their counterparts at their competitors' facilities to tell them about golden opportunities to hire qualified people. The net result was that every employee who was to have been laid off went to another job instead.

Case Analysis

Westinghouse had no legal obligation to find jobs for its laid-off employees. However, its executives firmly believed that they had an ethical obligation to help these employees and their families. A phrase often heard at Westinghouse in those days was "corporate responsibility," a principle everyone who worked at the company strongly believed in. Many companies do not operate that way today. Not all companies operated that way then either, and at least some of them no longer exist. Treating people fairly and acknowledging their inherent dignity is, besides being ethically responsible, sound business practice.

In summary, we can say that ethical standards contrast sharply with the notion that employees, clients, subcontractors, customers, and the people in the community at large are just numbers on a balance sheet. No matter how big it is, a company that regards people as nothing more than an opportunity for profit does not have much of a future in a free society.

Ethics in Technical Documents

Relating the Facts

Technical documents provide information. When the information will be used to make business decisions, technical writers or their supervisors are sometimes tempted to "cook" the information. Let's again consider the example given in the introduction to this chapter. Suppose you are writing a report that executives will use as a basis for deciding whether to go ahead with a project. Maybe you believe that thousands of jobs will be lost if you tell the unadorned truth. Are you ethically bound to tell the truth anyway? The short answer is *yes*.

Part of the long answer is that you must present the facts without trying to second-guess the executives who will make a decision based on your presentation. Naturally, if you have been directed to include recommendations, you must comply. If you include recommendations, make sure to do the following:

- Make your recommendations follow logically from the substance of the report.
- Make sure each recommendation includes a reference to relevant facts within the report.
- Acknowledge that your recommendations are based on the information you have at hand and that unknown or uncalculated factors beyond the scope of the report may negate them.

Another part of the long answer is that writers are free to use rhetorical devices, such as those discussed in Chapter 6, and that those devices affect emphasis. As long as writers do not abuse rhetoric, they behave ethically. However, when writers try to lead readers to false conclusions, they abuse the art of rhetoric. The line between use and abuse is sometimes hard to see, but with a little work, most writers can make sound ethical judgments.

As you discovered in Chapter 6, rhetoric is the art of using words to achieve a purpose. As human beings, we are much more than computers programmed to react to input. We come to conclusions about states of facts and then judge and respond to them in accordance with the freedom of human will. We may also argue in favor of our conclusions and against those of others. Although there is nothing inherently wrong in supporting our own thoughts and conclusions, using words and structures that put a particular "spin" on the facts can easily descend into an abuse of rhetoric. We will look at the uses and abuses of rhetoric in subsequent paragraphs. As we will see in those paragraphs and in the exercises that follow, the people who are fooled by honestly used rhetorical devices are, sometimes at least, those who want to be fooled.

Proposal writers often face difficult decisions in presenting information, since their goal is always the same: that of selling products or services. Experienced proposal writers know, however, that glossing over difficulties in meeting specifications or giving assurances instead of evidence and data doesn't fool anyone. Proposal evaluators are among the sharpest readers any technical writer will ever encounter. If there's "plenty of sizzle but not much steak," the competent proposal evaluator will know.

Layout and Presentation

Telling the truth, the whole truth, and nothing but the truth in a technical document may not be enough. As everyone knows, facts can be related in "large print" or "small print," as shown in the fictional direct-mail advertisement below:

YOU HAVE ALREADY WON ONE OF THE FOLLOWING PRIZES:

- Two weeks for two in Hawaii, all-expenses-paid
- A Sony Home Theatre
- A Designer Digital Watch by Forté
- An RCA Autofocus Camcorder

No salesperson will call. No gimmicks. Fill in the application form below and receive your prize, along with a free issue of *The Fubar Review,* a magazine for young professionals like you.

Publisher's Warranty: If you are not completely satisfied, simply check off the cancel box on the invoice that accompanies your free issue. Odds of winning trip (1 in 750,000), home theater (1 in 750,000), camcorder (1 in 750,000), digital watch (749,997 in 750,000).

Most people are familiar with the kind of "come-on" shown in this example. The reader is being asked to subscribe to a magazine that will come automatically forever unless it is canceled. But although a magazine subscription is what the advertisement is about, it doesn't specifically say that. Instead, it focuses the reader's attention on a "prize" he or she has "already won." A careful reading of the small type indicates that if everyone answered the advertisement, all but three of them would get digital watches as prizes. But even those who read the small type may think that the camcorder is the least expensive prize since it is mentioned last, and everyone knows that even the lowest priced camcorder is an expensive item.

Direct-mail experts estimate that about 7 percent of the people receiving this advertisement would subscribe, raising the magazine's circulation by nearly 50,000. The publisher would spend far more on the direct mailing than on the prizes, which more than likely would consist only of digital watches worth less than a dollar each anyway.

Is the advertisement ethical? May technical writers use similar techniques? Would the resulting documents be ethical?

As a basis for answering these questions, complete the quiz below and refer to the discussion that follows.

QUIZ DESIGNED BY DR. SAM DRAGGA OF TEXAS TECH UNIVERSITY[4]

1. A prospective employer asks job applicants for a one-page résumé. In order to include a little more information on your one page, you slightly decrease the type size and the leading (i.e., the horizontal space between the lines). Is this ethical?

   1. Completely ethical.   2. Mostly ethical.   3. Ethics uncertain.   4. Mostly unethical.   5. Completely unethical.

   Please explain:

2. You are preparing an annual report for the members of the American Wildlife Association. Included in the report is a pie chart displaying how contributions to the association are used. Each piece of the pie is labeled and its percentage is displayed. In order to de-emphasize the piece of the pie labeled "Administrative Costs," you color this piece green because cool colors make things look smaller. In order to emphasize the piece of the pie labeled "Wildlife Conservation Activities," you color this piece red because hot colors make things look bigger. Is this ethical?

   1. Completely ethical.   2. Mostly ethical.   3. Ethics uncertain.   4. Mostly unethical.   5. Completely unethical.

   Please explain:

3. You have been asked to design materials that will be used to recruit new employees. You decide to include photographs of the company's employees and its facilities. Your company has no disabled employees. You ask one of the employees to sit in a wheelchair for one of the photographs. Is this ethical?

   1. Completely ethical.   2. Mostly ethical.   3. Ethics uncertain.   4. Mostly unethical.   5. Completely unethical.

   Please explain:

4. You have been asked to evaluate a subordinate for possible promotion. In order to emphasize the employee's qualifications, you display these in a bulleted list. In order to de-emphasize the employee's deficiencies, you display these in a paragraph. Is this ethical?

   1. Completely ethical.   2. Mostly ethical.   3. Ethics uncertain.   4. Mostly unethical.   5. Completely unethical.

   Please explain:

[4] Dragga, Sam. From "Is This Ethical?" A Survey of Opinion on Principles and Practices of Document Design, *Technical Communication, Journal of the Society for Technical Communication,* Volume 43, Number 3, pp. 256–257. Used with permission from *Technical Communication,* published by the Society for Technical Communication, Arlington, Virginia.

5. A major client of your company has issued a request for proposals. The maximum length is 25 pages. You have written your proposal and it is 21 pages. You worry that you may be at a disadvantage if your proposal seems short. In order to make your proposal appear longer, you slightly increase the type size and the leading (i.e., the horizontal space between lines). Is this ethical?

   1. Completely ethical. 2. Mostly ethical. 3. Ethics uncertain. 4. Mostly unethical. 5. Completely unethical.

   Please explain:

6. You are preparing materials for potential investors, including a five-year profile of your company's sales figures. Your sales have steadily decreased every year for five years. You design a line graph to display your sales figures. You clearly label each year and the corresponding annual sales. In order to de-emphasize the decreasing sales, you reverse the chronology of the horizontal axis, from 1989, 1990, 1991, 1992, 1993 to 1993, 1992, 1991, 1990, 1989. This way the year with the lowest sales (1993) occurs first and the year with the highest sales (1989) occurs last. Thus the data line rises from left to right and gives the viewer a positive initial impression of your company. Is this ethical?

   1. Completely ethical. 2. Mostly ethical. 3. Ethics uncertain. 4. Mostly unethical. 5. Completely unethical.

   Please explain:

7. You are designing materials for your company's newest product. Included is a detailed explanation of the product's limited warranty. In order to emphasize that the product carries a warranty, you display the word "Warranty" in a large size of type, in upper- and lowercase letters, making the word as visible and readable as possible. In order to de-emphasize the details of the warranty, you display this information in smaller type and in all capital letters, making it more difficult to read and more likely to be skipped. Is this ethical?

   1. Completely ethical. 2. Mostly ethical. 3. Ethics uncertain. 4. Mostly unethical. 5. Completely unethical.

   Please explain:

Question 1. Cramming Information into a Résumé

Very few of the technical communicators and technical communication instructors surveyed[5] thought there was anything unethical in decreasing the type

---

[5] Ibid. p. 260.

size and line spacing to include more information in a one-page résumé. A more important question might be whether the resulting page layout would appear cluttered and make the author's résumé less appealing than those it would compete against.

Question 2. Using Color as a Rhetorical Device

Rhetoric is the art of using words and structures to persuade others to adopt a point of view. You have no doubt heard some people say, "Oh, that's nothing but rhetoric." They mean that the writer or speaker has used words and language structures dishonestly. Some people use rhetoric that way. They do so to inflame rather than to inform or to persuade with empty words rather than to convince with logic. But rhetoric is merely the art of using words well. A writer or speaker who uses words well does not automatically lie any more than a dull or verbally incompetent writer or speaker automatically tells the truth. In linear communication, then, rhetoric is a good thing as long as it doesn't replace logic and compromise the truth.

In spatial communication, the same principle holds true, and colors function in the same way that the positions of natural emphasis do (see Chapter 6). You would choose colors that enhance the message you wish to impart, just as you would go to a job interview wearing clothes that enhance your appearance. (Only 17.2 percent of Dr. Dragga's respondents thought this use of color was unethical.)

Question 3. Creating an Image

In Dr. Dragga's survey, 85.6 percent of respondents declared the fudged photograph to be unethical (76.1 percent thought it completely unethical).

To shed light on this situation, let's ask a few additional questions: What precisely is this company's policy with regard to handicapped workers? Does company policy exclude them? If so, the photograph has been designed to create a false image of the company, an unethical act to be sure.

But if the company is eager to employ handicapped workers and if its recruiters have been instructed to actively pursue them, might not this photograph further the effort by encouraging handicapped people to apply? The real question is whether the image presented is a true image of the company, not whether the company employs any handicapped workers at the moment.

However, in defense of those who think this act to be unethical no matter what the writer intended, let's be clear that fudging data to impart an image is the same as placing both feet on a slippery slope.

Question 4. Rhetorical Use or Abuse?

Fewer than 20 percent of those surveyed found using a bulleted list to display qualifications and a paragraph to discuss deficiencies to be an unethical act. The rhetorical methods chosen by the writer show a bias in favor of the employee, but the writer has not attempted to deceive the reader. Suppose, instead, that the writer had left out a discussion of deficiencies altogether. The evaluation would then appear objective, and it wouldn't matter whether the employee's qualifications were given in a bulleted list or a paragraph. But the evaluation would definitely be unethical because it would be incomplete.

Question 5. Using Graphic Design Changes to Match a Specified Document Length

Only 5 percent of the survey respondents thought that doctoring the graphic design to match a specified document length was unethical. As with Question 1, the real question here might be *Have you made a complete fool of yourself?* In the first place, the specified length was an *upper* limit, and no competent proposal evaluator would even bother to notice that your proposal didn't use up all the allotted space. Proposal evaluators look for more important things, such as technical merit and clarity of expression.

Question 6. Using an Unorthodox Graph

Two-thirds of the survey respondents labeled this practice unethical. Graph construction is a straightforward affair, and running the horizontal line in reverse is not an option (see Chapter 8). Therefore, this change has no parallel in any rhetorical practice. A writer who made such a clumsy attempt to deceive readers would most likely be fired for embarrassing the company.

Question 7. Graphic Design to Emphasize and De-emphasize

This question split the respondents at either extreme down the middle: 14.4 percent thought the practice completely ethical, and 14.4 percent thought the practice completely unethical. However, the "mostly unethical" voters outnumbered the "mostly ethical" ones by a large margin: 29.7 percent to 18.9 percent. The remaining 22.6 percent were in the "ethics uncertain" camp. So what is the answer?

This question posits a document situation similar in some respects to that of the direct-mail advertisement appearing as an example earlier. All the information is there in both cases. But while the direct-mail advertisement definitely crosses into unethical territory, the warranty explanation probably does not. The authors of the advertisement are not merely enthusiastic rhetoricians; they have set out to deceive their readers for the purpose of selling magazines.

The author of the warranty materials, on the other hand, is just trying to present the information in its best light. This author is saying, "Look, we have a warranty." Readers who want to find out more about it can struggle through the all-cap small print. The authors of the direct-mail advertisement are saying, "You already have won a prize, etc." Their purpose is to get people to buy magazine subscriptions without knowing what they're doing.

Does the distinction seem clear to you? If so, you understand the difference between rhetorical practice and plain deceit.

Exercises

16-1. Being Green[6]

1. After reading the following case very carefully, write a brief paper answering the following questions:

   a. Is there a solution to Lee's dilemma?

   b. Is it ethical to stay with SoftSolutions and help produce documentation that is environmentally harmful and wasteful of natural resources in spite of strong feelings that it's wrong to do so?

2. At the conclusion of the case, you will find responses written by professional technical writers. In order to gain access to more of the facts pertinent to this case, you should study the responses given by these writers before writing your own.

3. In composing your response, carefully consider both sides of the question (Lee's and Pat's). Set out the facts of the case, and determine what action Lee should take.

   Lee Adams, once the only writer in an aggressive start-up company, is now publications manager for that same company. The company, SoftSolutions,

[6] The case reprinted in this exercise originally appeared in Gear, John M. "Ethics Case: Being Green," *INTERCOM*, July/August 1995. It is used with permission from *INTERCOM*, published by the Society for Technical Communication, Arlington, Virginia.

is now a prosperous, profitable software firm about to issue its first public stock offering. And old friends are noticing that, instead of exhibiting his usual spark and energy, Lee seems more and more withdrawn.

Lee reports to Pat Kelly, who was brought in as director of marketing as part of the deal with the venture capital firm that helped keep SoftSolutions alive through the thin years. Pat's drive and sales experience have been instrumental in the company's growth. And Pat sees high-quality documentation as a critical and valuable part of the company's image.

But for Lee Adams that's less important than it once would have been. In the five years since Lee single-handedly wrote and assembled SoftSolutions' first product manual at a copy shop, things have changed. Now, Lee has a child and spends more free time on environmental issues than anything else. "It's only since we had Garrett that we really began to understand it all," Lee says. "Now we see the world differently — and that what we used to do without thinking isn't good. We can't afford not to think about the costs of what we're doing to the world."

And that explains Lee's all but total loss of interest in SoftSolutions: "We go about nearly everything wrong. And Pat's not willing to listen or change any of it."

In the spring, Lee approached Pat with three proposals:

1. Use only paper with low environmental impact: recycled, recyclable, unbleached papers and environmentally benign inks. This would mean doing without the slick papers and metallic glosses that Pat loves, not to mention paying a premium for each manual printed. It would also mean an end to the highly bleached, bright white paper that SoftSolutions began using when the broker who helped the company get venture capital said that its image wasn't businesslike enough.

2. Adopt a minimalist documentation strategy to reduce the volume of paper generated with each product. This would mean figuring out some way to give users high-quality instructions while reversing SoftSolutions' historical trend: The average weight of paper included with each package sold has doubled every two years.

3. Assume responsibility for ensuring that all the components in SoftSolutions' products are recycled. Lee envisions giving customers clear instructions for recycling every part of each product — even accepting original disks back to be overwritten with the latest program version after upgrades. Lee felt shocked by how fast Pat shot down each idea.

"That paper is our identity to our customers," Pat said. "SoftSolutions isn't in a warehouse incubator any more and we'd better not look like it.

"We've created — actually, Lee, you've created — a very valuable, consistent identity for SoftSolutions through our documents, and we're not about to throw it away so that we can see if there are three customers who know or care what a spotted owl is.

"And I'm not about to ask anyone to pay more so we can indulge ourselves on something like this. Our customers pay for value, and part of that value is a manual that reflects who we are — a high-tech, cutting-edge firm that puts out great products. Which, coincidentally, make a lot of money. And that means a lot for you after the stock offering. You've got even more options than I do.

"And where does this minimal manual stuff come from? Aren't you the one who's always going on about white space and typeface? Wide margins, detailed indexes, good examples — that's all you isn't it? Myself, gee, I used to think maybe we overdid it — I thought our tables of contents needed tables of contents — but you convinced me, and now I think you're right. Reviewers really love those fat manuals. And you're the one who always says we have to put the user first. If it's too long, people won't read it. But if there's not enough, they might not buy it.

"And recycling disks? Have you thought of the expense? We buy disks by the trainload, Lee. Think about it — we upgrade and suddenly we're supposed to get millions of disks back, clean them up, kill the viruses, write the upgrade in, and return them? We could just include a year's supply of Prozac with each package for less money and people would feel better besides.

"Lee, if you want to do the right thing for the world, give your money to the cause. But we're in business to make a profit, and we owe ourselves — and the shareholders we're gonna have — the best profit we can make. Because if we don't do that, we're gone. And then there won't be any money or stock options we can use to save the whales. No profit, no principles. It's as simple as that."

After mulling it over for a month, Lee was about to go see the partners who first founded SoftSolutions — but decided against it when Pat was made a vice president and invited to join the executive board. Now, Lee thinks about quitting every day except payday, when it seems clear that leaving SoftSolutions will just hurt Lee and his family and that SoftSolutions will go on without skipping a beat.

The foregoing ethics case appeared in *INTERCOM*, a publication of the Society for Technical Communication (STC), and the following are samples of the responses sent in by members of the Society.

From David N. Parker, STC, James River Chapter

Lee Adams' first problem is not doing his homework. The high-gloss papers he objects to may actually use less wood product than cheaper papers, and they are quite recyclable. These papers contain more filler than newsprint does, for example. Similarly, most mills today recycle their chemicals — including bleach chemicals. Finally, producing bright white recycled paper requires more chemical treatment than producing virgin white paper. Since paper is made primarily from farmed wood, I cannot agree that using paper is any more wasteful of natural resources than eating cereal for breakfast.

The accuracy of Lee Adams' concerns doesn't really matter, however. If he truly dislikes what he is doing — which is stated forcefully in the brief — he must leave.

His ethical problem is selling out his beliefs for salary and options. If he continues to do so, he will hurt himself and his family far more than will his (temporary) loss of income.

From Jack Greenfield, STC, Space Tech Chapter

An Open Letter to Lee Adams, Pubs Manager, SoftSolutions

Hey, Lee —

If you know so much, how come you don't know:

1. That recycled paper products are loaded with dioxins.

2. That the manufacture of recycled paper products releases more dioxins into the atmosphere than the burning of recycled paper products.

3. That bleaching reduces dioxin levels.

4. That metallic glosses chelate dioxins.

5. That the printing industry converted to soy-based inks years ago. (Even STC uses Soya inks in its publications.)

6. That the vision of top management around the world is to have minimalist documentation produced by the pubs managers. (That the mission of pubs managers around the world is to satisfy the top management vision of minimalist documentation.)

7. That your having reached your second level of incompetence will be communicated to you when top management brings in a pubs director over your head who will buy Soya inks and mandate minimalism in the pubs department. (This will be achieved in part by restoring you to your old writing job, as having a pubs director and a pubs manager would be overkill.)

8. That you reached your first level of incompetence when you lost touch with reality and stopped short of doing your real job as publications manager. (That your recycled paper cause was really a symptom of losing touch. Were I your manager and friend, I would speak to your wife about this concern of mine.)

9. That the best way to recycle disks is to make artsy-craftsy projects out of them.

10. That when you're collecting unemployment, having your wife sell your recycled disk projects at craft fairs and flea markets is a swell way of supplementing the family income.

16-2. *The Double-Blind Bind*[7]

1. After reading the following case very carefully, write a brief paper answering the following questions:

   a. Is Jennifer's desire to avoid "exclusionary" language an example of political correctness gone stratospheric, or is it a legitimate concern for technical communicators?

   b. Are Dr. Singh's resistance and arguments legitimate or reflections of the social comfort that has accompanied her success? If she herself were blind, would she have a different attitude?

2. In composing your response, carefully consider both sides of the question (Jennifer's and Dr. Singh's). Set out the facts of the case, and write your response.

3. At the conclusion of the case, you will find responses written by professional technical writers. You may wish to compare these responses with your own.

Jennifer LeLevier worked with Dr. Singh whenever she had the opportunity. Dr. Singh had become a mentor, a friend, and a model of what women could accomplish in a workplace traditionally dominated by men.

"Oh, she's brilliant," Jennifer had told her mother, with whom she still lived. "She's only in her early 30s and she's already a senior scientist at the lab. I bet she's director of the whole place in 10 years."

One of the qualities Jennifer liked best in Dr. Singh was her progressive attitude toward language. She was one of only two scientists who had agreed to let Jennifer use gender-neutral language when she edited their proposals and reports. For the first meeting, Jennifer had prepared an elaborate argument for replacing all those masculine singular personal pronouns. It was the same argument that had repeatedly evoked silence, jokes, or dismissal from all but one of the male scientists to whom she had proposed such changes. She was ready to launch the argument when Dr. Singh quickly and quietly said, "Yes, okay. Make the changes." And that was that.

---

[7] The case reprinted in this exercise originally appeared in Bryan, John G. "Ethics Case: The Double-Blind Bind," *INTERCOM*, November 1995. It is used with permission from *INTERCOM*, published by the Society for Technical Communication, Arlington, Virginia.

So when she decided to approach Dr. Singh with an even more progressive approach to the research proposal on which they were working, Jennifer expected a favorable hearing.

After lunch one Friday, she gathered her current draft in her arms and strode purposefully up the hallway to the senior scientist's office.

"We have to do something about this," she said, after dropping into the chair beside Dr. Singh's desk. "We have to come up with a suitable, nondiscriminatory substitute for 'double-blind.'"[8]

Dr. Singh removed her glasses and sat back in her chair. "I don't understand."

"Double-blind," Jennifer said, "discriminates against blind people. I know it isn't intended that way, but it marginalizes a segment of the population by taking their differentness and making it into an impersonal ..." She struggled for the right word. "... by making it something impersonal and mechanistic, it depersonalizes them."

Dr. Singh shook her head. "Surely a blind person doesn't take offense from that usage."

"Oh, yes," Jennifer said, surprised that Dr. Singh had not immediately recognized the gravity of the problem. "I'm quite sure they do. I was once told by a blind person that he didn't like the word 'blind' being used in that sort of way. We were students together and had a course in statistics. The professor kept talking about 'blind luck.' Every day, it seemed, he had some occasion to use that phrase: 'It's not blind luck.' Then he'd go on to talk about probability. Well, my blind friend hated that. 'What kind of luck is blind?' he would always ask me. He was convinced that the professor kept using that phrase because he didn't like him and didn't think he should be in college."

"Do you think that's what the professor intended?"

Jennifer shrugged. "I don't know. It didn't seem like it to me, but my friend said he could pick up hints of such attitudes, that he picked them up all the time from people. He said that he knew he made people feel uncomfortable and that they didn't want him around."

"It sounds to me as if your friend had a special difficulty with that professor and that maybe his objection to 'blind luck' came more from that troubled relationship than from the phrase itself."

"But we still have to be sensitive, don't we? Just because he had a problem with that professor, does that mean we shouldn't be sensitive to things that he might find hurtful?"

"No, but isn't there some limit to how far we must go in dancing around the possibility of offending someone? What if I objected when someone said 'holy cow'? I am Hindu and find that particular expression slightly objectionable because it implicitly ridicules a tradition of my religion. But if you

---

[8] The term "double-blind" is used to refer to scientific studies in which neither the investigators nor their subjects know which of the subjects are acting as controls.

asked 10 Americans about the origins of that expression, I bet nine of them would make no connection to Hindu. So I take no offense. None is intended."

Jennifer sat up in her chair and leaned forward. "Oh, I've said 'holy cow' before and I had no idea." She hesitated before continuing. "But, of course, that's right. I see it now. And you should be offended. It's not right."

"Jennifer, you are missing my point. If I heard you say 'holy cow' I wouldn't be offended because I know you intended no offense. You see? Besides, I would think that you, as a professional writer, as a person who knows and loves language so well, would find it painful to see language stripped of all it metaphors, of all its vivid terms and phrases."

Jennifer didn't much like the personal tone that the conversation had just taken. She knew all about metaphors, and she did have a special love for language, she thought. She didn't need to have those things pointed out to her. "But vivid language can be very hurtful. The first time I told my boyfriend about you and said you were Indian, he asked whether you were a 'red-dot' Indian or a 'woo-woo' Indian. I nearly hit him."

Dr. Singh smiled. "Well, Jennifer, I appreciate your impulse to defend my honor, but I'm probably less sensitive than you suspect."

"Ah, but is everyone? Would none of your Indian friends be offended?"

"I don't know. I do know that you are the expert on language and I am only a dilettante, but it seems to me that we need to be careful in changing our language, careful in reducing it to the lowest common denominator in an effort to make it palatable to every person worldwide. Can I not say that I am in a black mood because that reinforces the connotation of black as negative and so is racist? Can I not tell my children to study hard so they don't end up in dead-end jobs because referring to the entirely honorable job of washing dishes is classist? If I use the popular Americanism 'I hear you' to mean 'I understand you,' does that offend and exclude deaf people? I'm not suggesting that we permit grossly offensive language, such as racial epithets and sexist obscenities, but metaphorical language — like 'double-blind' — is so succinct, so precise, and usually so well intended and well understood, that it pains me to think we must abandon it. What else could I use to describe a double-blind study?"

The foregoing ethics case appeared in *INTERCOM,* a publication of the Society for Technical Communication (STC), and the following are samples of the responses sent in by members of the Society.

From Andrea Holloway, STC, Mid-South Chapter

Having finished reading "Ethics Case: The Double-Blind Bind," I felt that I must write this letter. The belief that this young professional holds, that all seemingly offensive language must be softened (censored), is becoming increasingly popular. As a technical communicator and one who truly appreciates the language, I find this trend deeply disturbing.

Censoring our language — even with the best of intentions — is harmful because it covers up the true meaning of words and tampers with free speech. Many of those who advocate changing our language to cater to the needs of everyone are missing the point. The proponents of politically correct extremism argue that making our language more accepting helps to support a more diverse society. This assertion couldn't be farther from the truth. Instead, it oppresses the writer and neuters the language, rendering it bland and lifeless. We are reduced to some politically correct common denominator, which doesn't actually exist. This kind of censorship does not actually deal with societal problems; it simply places a Band-Aid over a sucking chest wound, if you will.

Censorship has not ever been, and will never be, the solution to society's problems.

From Paul S. McKelvey, STC, Austin Chapter

Jennifer is far too sensitive for her own good. Even when her boss says that she takes no offense at something Jennifer has said, Jennifer frets.

As a practical matter, much "political correctness" slips into the realm of the absurd. Its advocates tremble whenever they believe they might offend someone. Note that Jennifer sets the standard for what is offensive. Her boss, despite protest that she is not offended, is not allowed to participate in that decision. Jennifer will be offended for her boss, regardless.

This is silly.

Finally, some people will be offended, no matter what is done. They diligently seek out offense in the nooks and crannies of the land and cry out mightily whenever they find it. For our part, we can listen patiently, hand them a cup of peppermint tea, murmur platitudes, and get on with this exhilarating business we call life.

From Neil Lineberger, STC, Atlanta Chapter

Jennifer is to be commended for her desire to eliminate all potentially offensive words and phrases from English, but she is facing an endless and futile effort. Dr. Singh's position is more realistic.

Our language is a collection of ideas that now might be viewed as offensive. Woman originally meant something like "a man with a womb," a characterization many would not accept today. But we still use the word because it fulfills a necessary purpose. And my sightless friends often say, "I'll see you tomorrow," or "Let's see if this works" without a second thought. "See" in those contexts is an effective metaphor. If we thought hard enough, we could find offense in almost any language.

At the same time, I sometimes avoid language I personally consider acceptable because it will steal attention from my point (just as I sometimes use jargon I don't like because my specialized audience will understand it). One of the first principles of good communication is to adapt the message to the audience.

The evolution of ideas and language will eventually solve this argument, and our culture will move on to other concerns. In the meantime, communication professionals must find a workable balance between practicality and sensitivity. When we know a word or phrase is patently offensive to our audience, then we should look for substitutes.

16-3. The Need To Know[9]

1. Read the following case very carefully and write a brief paper explaining Chandala Liggitt's options and recommending the action she should take.

2. In your response, answer the following questions:

   a. Did Chandala Liggitt's glimpse through the "need-to-know" door oblige her to try to find out more? If she has such an obligation, would her financial dependency on her employer and her responsibility to her children relieve her of her responsibility to find out the truth?

   b. Does the apparent goal of the project (she is not absolutely sure of it) affect her responsibility?

3. At the conclusion of the case, you will find responses written by professional technical writers. You may wish to compare these responses with your own.

After a four-year absence from the technical writing job market, Chandala Liggitt sought to return two days after her husband abandoned her and her two children. With little money in savings — and mortgage, credit-card, and car payments to contend with — she was desperate to find a job.

Three weeks later, after numerous failed attempts, Liggitt landed a job as a technical writer with a small medical equipment research lab. The lab's CEO informed her that their researchers were developing a device that would revolutionize medicine, that their client had given them 16 months to deliver the device, and that the lab operated on a strict "need-to-know" basis.

---

[9] The case reprinted in this exercise originally appeared in Bryan, John G. Summary of the "Ethics Case: The Need to Know," *INTERCOM*, June/July 1996. It is used with permission from *INTERCOM*, published by the Society for Technical Communication, Arlington, Virginia.

Even after 10 months on the job, Liggitt still did not have comprehensive knowledge of the project, although the bits and pieces she saw indicated that the device was intended to provide an improved means of detecting cancer. As the deadline for the project's completion approached, the mood of the project team darkened. A month before the deadline, the CEO and project manager returned after a brief absence and announced that the project funding would be extended for a year and expanded to fund new clinical trials.

Later that day, Winston Williams, the project manager, confided to Liggitt that the researchers would never complete the project. When Liggitt inquired further, Williams shook his head. "You don't want to know, Chandala," he said. "In another year, you'll want to be able to look back and say, 'I didn't know a thing.'"

The foregoing ethics case summary appeared in *INTERCOM*, a publication of the Society for Technical Communication (STC), and the following are samples of the responses sent in by members of the Society.

From Deborah Snavely, STC, Silicon Valley Chapter[10]

> "To what extent does that brief glimpse through the need-to-know door obligate Liggitt to pursue the truth?"

Liggitt "wanted to know" what was outside her cubicle. She's on the brink now, with Williams's remark — she can ask more and lose her apparently valuable ignorance, or she can turn away and retain it. She's already opened the conversation and continued into gray territory ... she'd be wise to stop talking and say goodnight *now*.

Alternatively, she can ask him outright whether there's anything illegal happening with the project; the downside of asking is the possibility that he'll say yes, and her ethical obligations to (at least) get out then become critical.

> "... financial dependency ... and ... responsibility toward her children relieve her of such obligation?"

Her responsibilities to her kids equally demand that she not put herself in jeopardy of possible legal proceedings or job blacklisting. How employable would she be if some later investigation or business fiasco muddies her name along with that of the project? Money considerations affect her actions, preventing her from just giving notice immediately, but she should plan for a graceful exit, whether now or soon or even at project "end." In my opinion, the sooner the better.

---

[10] This case is given in summary form, as it appeared in the June/July 1996 issue of *INTERCOM*. Deborah Snavely's quotations refer to the full-length version of the case, which appeared in the April issue of *INTERCOM*.

"Does the ... goal ... change her responsibility for finding and exposing the truth?"

The goal is irrelevant; the end does not justify the means. However, medical investigations have been known to produce useful results through serendipity. It's possible that even a failed project could have worthwhile output. But Liggitt cannot count on that.

Liggitt cannot afford to become a whistle-blower from within. Williams's clear certainty (now that she's learned of it by her own actions) that there will be questions asked later and his statement, "It's a bust," implies possible fraud, given the continued funding. "We'll never produce this thing," he said. With her family responsibilities, Liggitt should look for the escape hatch first, but consider confronting management with her doubts on her way out.

From Robert Lessman, STC, Silicon Valley Chapter

I have personally known writers who were in similar situations to Chandala. In Sun Tzu's book, *The Art of War,* his advice was that the battle not fought is the battle not lost. Chandala is caught in a no-win situation. She should brush up her résumé and find other work. She can't afford to leave until she has a firm offer, but she must leave soon to keep her integrity intact.

Most writers have a driving quality that makes them dig for the facts they need as a matter of pride. Remaining pleasantly ignorant would degrade and demoralize any writer. She does not have the resources to change a shady situation, but she might have supportive resources on a new job.

When all hell breaks loose at that company, the integrity of everyone who stayed to the end will be under suspicion. For the sake of her children and her own career, she can't afford the stigma that will land on those who remain.

From Jon Russell, STC, Washington, D.C. Chapter

Chandala Liggitt's quandary about the cloak of mystery shrouding the "cancer detecting device" at the small medical equipment research lab provides her with a daunting challenge. Does Ms. Liggitt keep her mouth shut and take the money and run? Or does she respond to her conscience and help uncover foul play that may harm the general public? (Though the article suggests that the project will go bust, it leaves open the possibility that another lab could pick up the product and produce it.)

This case dredges up the tired, though continually nagging, paradox: Do we condemn the man or woman who steals a loaf of bread for his or her starving children? After all, Chandala was doing all she could despite a wanderlusting and irresponsible husband who lurked in the background of her life, demoralizing her and possibly forcing her to prioritize any "crusades for justice."

Therefore, the question is, does Ms. Liggitt fight both for her children and the general public, jeopardizing the income that cares for herself and her children? My first inclination is to advise her not to bother with uncovering the truth of the lab. She should prepare her résumé and look for a job during her lunch hour. In short, get out of there as fast as she can — put her conscience on hold and hope it goes away.

However, it is conceivable that she can protect her well-being, her children, and the general public by proceeding as follows:

- As suggested earlier, Ms. Liggitt should prepare her résumé and look for another job.

- As she is searching for another job, she can make use of her remaining time at the lab by talking with employees like Williams to learn facts that might shed light on the mystery shrouding the lab — possibly uncovering facts leading to potentially dangerous design plans that may adversely affect the general public.

- If she finds factual evidence that could lead to harming the general public, the best and most efficient means to ferret out the truth is to deliver her facts to a reputable newspaper editor who can assign a reporter to investigate the lab. Also, she should notify the police.

Though Ms. Liggitt may be taking some extreme chances in this circumstance, she did the work that was required of her conscience. She has remained courageous and bothered to get involved. No doubt she feels better about herself, too. Who knows how many lives she may have saved?

Ms. Liggitt's ultimate victory was achieved by seeking to protect the financial well-being of herself and her children, and secondarily the well-being of the general public.

I haven't come close to offering even a semblance of a solution to the paradox: Do we condemn a man or woman for stealing bread for his or her starving children? The paradox remains triumphant yet another day.

A
Self-Directed
Course
In English
Grammar
and Usage

Including Answers to Exercises

Grammar is a science rooted in logic. Therefore, grammar must be approached phenomenologically, as a study of how language structures relate to human experience as a whole, rather than as an empirical investigation that restricts itself to blunt observation.

This Self-Directed Course takes such an approach. The rules of usage are given in unequivocal terms; exceptions to the rules are not treated as rules in themselves. As a technical writer, your first obligation is to communicate clearly and concisely. If breaking a rule will help you to achieve some stylistic purpose, you will know what rule you are breaking. Until you reach that level of competence, however, you should regard the usage rules contained herein as absolute.

The Parts of Speech

At one time or another, you may have learned the parts of speech, of which there are eight:

1. Noun	3. Verb	5. Adverb	7. Conjunction
2. Pronoun	4. Adjective	6. Preposition	8. Interjection

A noun names a person, place, or thing. *Ralph, stadium, Vancouver, shoe,* and *computer* are examples of nouns. Some words can act as more than one part of speech. Function in the sentence determines what part these words play. Consider the following example:

> We didn't **practice** yesterday. The **practice** was called off when the **practice** ball couldn't be found.

The word *practice* is used three times in the example. It is successively a verb, a noun, and an adjective.

A pronoun is a single word that takes the place of a noun. *I, you, he, she, they, us, which, that, these, this,* and *those* are examples of pronouns.

A verb is a word that denotes the action in the sentence. The action can be physical or mental, as illustrated in the following examples:

Sally **kicked** a field goal. (physical action)

Sally **likes** football. (mental action)

The verb *to be* is a special verb that denotes existence. In the sense that existence is a physical phenomenon, this verb fits the physical action category. The verb *to be* is often used to link a subject with its complement, as in the following example:

Sally **is** a good football player.

An adjective modifies a noun. *Beautiful, angry, gracious, great,* and *small* are examples of adjectives.

An adverb modifies a verb, an adjective, or another adverb. Some adjectives can be converted into adverbs by adding *ly* to them. Some examples of adverbs are *beautifully, graciously, pridefully, never, too,* and *madly.*

Prepositions introduce an object. Together with its object, which may also be accompanied by modifiers, the preposition forms a prepositional phrase. Prepositional phrases nearly always act as either adjectives or adverbs. Some examples of prepositions are *to, for, in, on, under, above, of,* and *from.*

Conjunctions connect words and groups of words. They also tell something about the relationships between the words they connect, and to that extent they can have an adverbial function as well as a connective one. Some of the conjunctions are *and, but, for, or, nor, yet, so, while, because, although, before,* and *after.*

The interjection is not related to any specific part of the sentence and has, therefore, little importance in the study of sentence structure. It is an exclamation or expression of emotion. For example:

Oh, how should I know?

Ouch, that hurts!

Exercises

A-1. Identifying Nouns and Verbs

In the following sentences, underscore the nouns and place parentheses around the verbs.

1. Students learn something new every time they follow the steps in the writing process.

2. When writing a first draft, you should work fast.

3. You can correct misspelled words when you write a second draft.

4. You may correct grammatical errors in a first draft, too.

5. A misspelled word or a heavy style only distracts readers.

A-2. *Identifying Adjectives and Adverbs*

In the following sentences, underscore the adjectives and place parentheses around the adverbs.

1. Students learn something new every time they follow the steps in the writing process.

2. When writing a first draft, you should work fast.

3. You can correct misspelled words when you write a second draft.

4. You may correct grammatical errors in a first draft, too.

5. A misspelled word or a heavy style only distracts readers.

A-3. *Identifying Pronouns, Prepositions, and Conjunctions*

In the following sentences, underscore the pronouns, place parentheses around the prepositions, and bracket the conjunctions.

1. Students learn something new every time they follow the steps in the writing process.

2. When writing a first draft, you should work fast.

3. You can correct misspelled words when you write a second draft.

4. You may correct grammatical errors in a first draft, too.

5. A misspelled word or a heavy style only distracts readers.

The Characteristics of Verbs

Verbs have five characteristics: *person, number, tense, voice,* and *mood.*

Person

The three persons are *first, second,* and *third.* Some verbs change form for each person. For example, the verb *to be* is conjugated as follows:

I am (first person)
You are (second person)
He, she, it is (third person)

Such verbs are called irregular. But even regular verbs change form in the present tense for the third-person singular:

I kick (first person)
You kick (second person)
He, she, it kicks (third person)

Number

Verb forms also reflect the singular or plural nature of their subjects. Again, the verb *to be* shows how the forms change for the plural pronouns:

We are (first person)
You are (second person)
They are (third person)

Tense

Verbs also change to reflect the time indicated:

Present	Past	Future
I am	I was	I will be
You are	You were	You will be
He, she, it is	He, she, it was	He, she, it will be
We are	We were	We will be
You are	You were	You will be
They are	They were	They will be

Present Perfect	Past Perfect	Future Perfect
I have been	I had been	I will have been
You have been	You had been	You will have been
He, she, it has been	He, she, it had been	He, she, it will have been
We have been	We had been	We will have been
You have been	You had been	You will have been
They have been	They had been	They will have been

Notice that for the perfect tenses the past participle of the verb is used as the principal verb and that auxiliary words are used to establish the tense.

Voice

Voice has to do with the flow of action in a sentence. The action flows from the subject in active-voice constructions, whereas the action flows back toward the subject in a passive-voice construction. A simple example illustrates this flow:

The soloist played a Mozart Concerto. (active voice)

A Mozart concerto was played by the soloist. (passive voice)

The subject in the active-voice construction performs the action described by the verb, whereas the subject in the passive-voice construction is the recipient, result, or product of the action.

Mood

English verbs exhibit three moods: *indicative, imperative,* and *subjunctive.* The indicative mood, sometimes also called the declarative mood, makes a statement:

The mountains in Banff National Park are very beautiful.

The imperative mood commands action:

Go to Banff National Park, and see the beautiful mountains.

The subjunctive mood indicates condition or desire:

If you go to Banff National Park, you will see the beautiful mountains.

If I were able to live in Banff National Park, I would enjoy the beauty of the mountains during every season of the year.

Notice that in the conditional clause in the final example, the verb form *were* is used in place of the form *was*.

A-4. Conjugating Verbs

Conjugate the following verbs in their present, past, and future forms.

1. to fight   2. to print

A-5. Conjugating Verbs

Conjugate the following verbs in their present perfect, past perfect, and future perfect forms.

1. to build   2. to take

A-6. Identifying Verb Voice

Identify each of the following sentences as having an active-voice or a passive-voice main verb by placing an A or a P in each space.

1. _____ Good technical writers have mastered grammar.

2. _____ The essential message emerges during the rewriting step.

3. _____ Not all technical documents are governed by specifications.

4. _____ The product or service requested is specified by the RFP.

5. _____ Good writing of any kind is hard work.

A-7. Identifying Verb Mood

Identify each of the following sentences as having an indicative, imperative, or subjunctive mood by placing IND, IMP, or SUB in each space.

1. _____ All technical writers should have mastered grammar.

2. _____ Make the essential message emerge during the rewriting step.

3. _____ Technical writers should know how to write to specifications.

4. _____ To write a winning proposal, carefully follow the RFP.

5. _____ Instructional steps are written in second-person, imperative mood.

The Elements of the Sentence

At the sentence level, the eight parts of speech are reduced to four: noun, verb, adjective, and adverb. Each of these four main parts of sentences can be either single words or groups of words. Groups of words are called clauses and phrases.

Clauses

A clause is a group of related words that contains a subject and a verb. An independent clause is one that can logically stand alone, and a dependent clause is one that cannot do so. The following examples illustrate the distinction.

> The computer has changed our way of life. (independent clause)

> **Although the computer has changed our way of life,** it has had little effect in undeveloped parts of the world. (dependent clause)

By placing a subordinate conjunction in front of our independent clause, we turned it into a dependent clause. Dependent clauses act as nouns, adjectives, and adverbs. In the foregoing example, the clause appearing in boldface type is just a nine-word adverb modifying the verb *has had* in the main (independent) clause of the sentence. The following examples show how clauses can act as nouns and adjectives:

> **That the computer has changed our way of life** is indisputable.

> The computer **that is best for desktop publishing** is the Macintosh.

The first example has a dependent clause (noun) as the subject of the sentence, and the second example shows a dependent clause (adjective) modifying the subject of the sentence.

Phrases

The phrase is a group of related words that does not have a subject and a verb. Phrases can act as nouns, verbs, adjectives, or adverbs.

A noun phrase is composed of a noun and its modifiers, as in the following example:

The tall, gray-haired, distinguished gentleman is the Freedonian Ambassador.

A gerund phrase is a special kind of a noun phrase, formed from the present participle of a verb and an object:

Running for public office requires the sacrifice of time and money.

An infinitive phrase can also act as a noun phrase:

To run for public office requires time and money.

A verb phrase is composed of a verb and its auxiliaries:

The Freedonian Ambassador **has been given** a warm welcome on the occasion of all his visits to this country.

Adjective phrases are formed by prepositions and their objects (and modifiers), by participles of verbs and their objects (and modifiers), or by the word *to* and a verb (infinitive phrase). Some examples follow:

The woman **in the blue suit** is the chief executive officer of the company. (prepositional phrase, adjective)

The woman **wearing the blue suit** is the chief executive officer of the company. (participial phrase, adjective)

Fearing the worst, the crowd became restive. (participial phrase, adjective)

The reasons **to vote for our candidate** are obvious. (infinitive phrase, adjective)

Adverb phrases are formed by prepositions and their objects and modifiers, or by the word *to* and a verb (infinitive phrase), as illustrated in the following examples:

Everybody arrived **on time.** (prepositional phrase, adverb)

Everybody arrived **in good spirits.** (prepositional phrase, adverb)

Everybody arrived **to congratulate Harry.** (infinitive phrase, adverb)

A-8. *Identifying Clauses*

In following sentences, underline the independent clauses and place dependent clauses in parentheses.

1. Many technical documents are associated with getting a product to market.

2. Major technical documents are almost never written by one person.

3. Abbreviation standards vary from one technical area to another.

4. Many people mistakenly believe that a proposal is a kind of report.

5. Writing a good résumé is easy if you remember a few principles.

A-9. *Identifying Phrases*

Identify the underscored phrase in each of the following sentences by placing N (noun phrase), ADJ (adjective phrase), V (verb phrase), or ADV (adverb phrase) in the spaces provided.

1. _____ All technical writers should have mastered grammar.

2. _____ Make the essential message emerge during the rewriting step.

3. _____ Technical writers should know how to write to specifications.

4. _____ To write a winning proposal, carefully follow the RFP.

5. _____ Instructional steps are written in second-person, imperative mood.

Sentence Structure

Let us begin the study of sentence structure by reviewing what we have thus far discovered.

- Sentences can be made up of single words, phrases, and clauses.
- Single words can function as nouns, pronouns, verbs, adjectives, adverbs, prepositions, conjunctions, or interjections.
- A phrase is a group of related words without a subject and verb.
- Phrases can act as nouns, verbs, adjectives, and adverbs.

- Noun phrases always act as nouns, and verb phrases always act as verbs.
- Infinitive phrases can act as nouns, adjectives, or adverbs.
- Prepositional phrases usually act only as adjectives or adverbs.
- Participial phrases can act only as adjectives.
- A clause is a group of related words containing a subject and a verb.
- Clauses that can stand alone are called independent clauses, and clauses that cannot are called dependent clauses.
- Dependent clauses can act as nouns, adjectives, or adverbs.

The Four Sentence Types

The basis of the English sentence is the subject-verb-object pattern; that is, something does something. Many variations on this pattern can occur, as we will see in the subsequent discussion.

SIMPLE SENTENCES

Simple sentences have one independent clause and no dependent clauses. Simple sentences are not always short, however, as illustrated in the following example of a long simple sentence:

> Sentence structure can be learned by anyone willing to spend a little time in learning terminology and willing to apply the principles of patient analysis in the pursuit of an understanding of sentence elements and their relationships.

COMPLEX SENTENCES

Complex sentences contain only one independent clause and at least one dependent clause. The following example illustrates a complex sentence.

> Sentence structure can be learned by anyone who is willing to devote a little time and effort.

Note that the addition of *who* and *is* to *willing* changes the character of this sentence from that of the preceding example, which used *willing* as a participle to form an adjective phrase.

COMPOUND SENTENCES

Compound sentences contain two or more independent clauses joined by one of the coordinate conjunctions and separated by a comma, or they contain two or more independent clauses separated by a semicolon. The coordinate conjunctions are *and, or, nor, but,* and *for.* Many writers now also include *yet* and *so* on the list of acceptable coordinate conjunctions. Here are two examples of the compound sentence:

> Nobody would expect a surgeon to operate without first studying human anatomy, but some teachers expect a student to write without first learning grammar.

> Nobody would expect a surgeon to operate without first studying human anatomy; that some teachers expect a student to write without first learning grammar makes no sense.

The second example requires at least a semicolon, since no connector of any kind is present.

COMPOUND-COMPLEX SENTENCES

Compound-complex sentences have at least two independent clauses and at least one dependent clause:

> If a surgeon who had no knowledge of human anatomy regularly lost patients on the operating table, everyone would know the reason, but few people seem to know the reason that those college graduates who have no knowledge of grammar write poorly.

Sentence Dynamics

Let us now examine some sentences to see how the system works:

> The trouble that we have with English stems most often from our inability to identify individual sentence elements.

The first step in analyzing a sentence is to find the main idea or ideas (independent clauses). In the foregoing example, we find one independent clause, the subject and verb of which are *trouble stems.* The next step is to look for other clauses, and we find *that we have with English.* This clause acts as an adjective modifying *trouble.* These are the only two clauses, and since we know that one of them is independent and the other is dependent, we know also that we have a complex sentence.

Next, we look for phrases, and we find *from our inability* and *to identify individual sentence elements.* The prepositional phrase *from our inability* relates directly to the verb *stems,* and we thus see that this phrase is an adverb. The infinitive phrase *to identify individual sentence elements* relates to the noun *inability,* which acts as the object within our prepositional phrase. Therefore, we know that this infinitive phrase is an adjective.

The only three words unaccounted for are *the, most,* and *often. The* is an adjective modifying *trouble,* and *often* is an adverb modifying the verb *stems.* The remaining word *most* is an adverb modifying the other adverb *often.*

Note: The word *the* is also referred to as the definite article. It is called "definite" because it defines or exactly identifies the noun it modifies. The words *a* and *an* are called indefinite articles. The definite and indefinite articles are considered adjectives.

A-10. *Identifying Sentence Types*

Identify each of the following sentences by placing S (simple), C (compound), CX (complex), or CCX (compound-complex) in the spaces provided.

1. _____ All warranty work is done on a local basis.

2. _____ Dealers who have technicians on staff complete the warranty work in their own shops.

3. _____ Dealers assess the appropriateness of each claim, and their technicians carry out the work.

4. _____ We should issue a warranty booklet with each instrument sold.

5. _____ The problem has a four-part solution.

Answers to Part A Exercises

A-1. *Identifying Nouns and Verbs*

(p. 333)

1. <u>Students</u> (learn) something new every <u>time</u> they (follow) the <u>steps</u> in the writing <u>process</u>.

2. When writing a first <u>draft</u>, you (should work) fast.

3. You (can correct) misspelled <u>words</u> when you (write) a second <u>draft</u>.

4. You (may correct) grammatical <u>errors</u> in a first <u>draft</u>, too.

5. A misspelled <u>word</u> or a heavy <u>style</u> only (distracts) <u>readers</u>.

A-2. *Identifying Adjectives and Adverbs*

(p. 334)

1. Students learn something <u>new</u> <u>every</u> time they follow <u>the</u> steps in <u>the</u> <u>writing</u> process.

2. When writing <u>a</u> <u>first</u> draft, you should work (fast).

3. You can correct <u>misspelled</u> words when you write <u>a</u> <u>second</u> draft.

4. You may correct <u>grammatical</u> errors in <u>a</u> <u>first</u> draft, (too).

5. A <u>misspelled</u> word or <u>a</u> <u>heavy</u> style (only) distracts readers.

A-3. *Identifying Pronouns, Prepositions, and Conjunctions*

(p. 334)

1. Students learn <u>something</u> new every time <u>they</u> follow the steps (in) the writing process.

2. [When] writing a first draft, <u>you</u> should work fast.

3. <u>You</u> can correct misspelled words [when] <u>you</u> write a second draft.

4. <u>You</u> may correct grammatical errors (in) a first draft, too.

5. A misspelled word [or] a heavy style only distracts readers.

A-4. *Conjugating Verbs*

(p. 337)

1. to fight

Present	Past	Future
I fight	I fought	I will fight
You fight	You fought	You will fight
He, she, it fights	He, she, it fought	He, she, it will fight
We fight	We fought	We will fight
You fight	You fought	You will fight
They fight	They fought	They will fight

2. to print

Present	Past	Future
I print	I printed	I will print
You print	You printed	You will print
He, she, it prints	He, she, it printed	He, she, it will print
We print	We printed	We will print
You print	You printed	You will print
They print	They printed	They will print

A-5. Conjugating Verbs

(p. 337)

1. to build

Present Perfect	Past Perfect	Future Perfect
I have built	I had built	I will have built
You have built	You had built	You will have built
He, she, it has built	He, she, it had built	He, she, it will have built
We have built	We had built	We will have built
You have built	You had built	You will have built
They have built	They had built	They will have built

2. to take

Present Perfect	Past Perfect	Future Perfect
I have taken	I had taken	I will have taken
You have taken	You had taken	You will have taken
He, she, it has taken	He, she, it had taken	He, she, it will have taken
We have taken	We had taken	We will have taken
You have taken	You had taken	You will have taken
They have taken	They had taken	They will have taken

A-6. Identifying Verb Voice

(p. 337)

1. A;   2. A;   3. P;   4. P;   5. A

A-7. Identifying Verb Mood

(p. 337)

1. SUB;   2. IMP;   3. SUB;   4. IMP;   5. IND

A-8. Identifying Clauses

(p. 340)

1. <u>Many technical documents are associated with getting a product to market</u>.

2. <u>Major technical documents are almost never written by one person</u>.

3. <u>Abbreviation standards vary from one technical area to another</u>.

4. <u>Many people mistakenly believe</u> (that a proposal is a kind of report).

5. <u>Writing a good résumé is easy</u> (if you remember a few principles).

A-9. Identifying Phrases

(p. 340)

1. N;   2. ADV;   3. V;   4. ADV;   5. V

A-10. Identifying Sentence Types

(p. 343)

1. S;   2. CX;   3. C;   4. S;   5. S

An Introduction to English Usage

Taboos surrounding English usage are so prevalent that they are the only "rules" many high school graduates know. Here are a few examples of usage taboos: never begin a sentence with *and* or *because;* never put a comma in front of *and* or *or;* never say *Jack and me;* and so on. It is no wonder that many students think English usage rules are arbitrary and hate learning them. In fact, rebelling against the taboos that have replaced the genuine usage rules is a necessary first step in mastering English usage.

Technical writers must not be satisfied with loose interpretations of English usage questions. The rules given herein are grounded in logic. When you decide to break a rule, be sure you know what the rule is and what stylistic effect you wish to achieve by breaking it, bearing in mind that what may appear as an exception represents a stylistic choice and not a grammatical option.

You may wish to make reference to the Grammar Fundamentals in Part A when the terms used in a rule are unclear to you. If you need further help in learning English usage, pick up a copy of *Plain English: A Guide to Standard Usage and Clear Writing,* Prentice-Hall Canada, or a copy of Wilson Follet's *Modern American Usage, A Guide,* Grosset & Dunlap.

Punctuation

Comma Rules

The comma is the mark used to show small separations in continuity of thought. The following rules can be applied absolutely. There are no exceptions.

RULE 1. Always place a comma before a conjunction that joins independent clauses. Do not make exceptions for short clauses.

> The Quality Control Department will establish the test procedures, and the Receiving Department will carry out the tests.

Note: In sentences structured as is the foregoing example, the comma will always be a correct choice. As a writer, however, you have the stylistic option of adding emphasis to the second half of such a sentence by using a semicolon in place of the comma. The sentence would then read as follows:

> The Quality Control Department will establish the test procedures; and the Receiving Department will carry out the tests.

Using a period in place of the semicolon would add even more importance to the second statement.

> The Quality Control Department will establish the test procedures. And the Receiving Department will carry out the tests.

RULE 2. Always place a comma after an introductory sentence element. Do not distinguish between words, phrases, or clauses.

> Eventually, the drawings were completed. (word)

> After completion, the drawings were delivered to the shop floor. (phrase)

> After the drawings were completed, they were delivered to the shop floor. (clause)

RULE 3. Always set off non-restrictive words, phrases, or clauses by placing commas around them. Such words include those indicating direct address and all other "parenthetic" elements.

> The three preproduction units, AN/SSF-40s, were subjected to environmental testing. (word)

> Three AN/SSF-40s, all preproduction units, were subjected to environmental testing. (phrase)

Three AN/SSF-40s, which were preproduction units, were subjected to environmental testing. (clause)

RULE 4. When items in a series do not contain internal commas, separate them by placing a comma after each item except the last. Please note that the rule does not say "next to last."

The simulated environments to which the units were subjected were altitude, temperature, humidity, and shock.

The comma's function is to separate the items so that readers will know where one item ends and the next one begins. Therefore, never omit the final serial comma. Putting it in will *never* cause confusion.

RULE 5. Place a comma between coordinate adjectives.

The long, involved test program led to a net loss for the quarter.

Adjectives are coordinate when they have a more or less equal relationship to the noun. You can test for the coordinate nature of adjectives in tandem by mentally placing an *and* between them. If the *and* sounds right, the adjectives are coordinate; if it does not sound right, they are not, and you should not place a comma between them.

The woman waiting in the outer room is carrying a brown leather briefcase.

The briefcase is brown, and it is leather; however, although both adjectives refer to the briefcase, they do not bear an equal relationship to it. Put another way, the briefcase is brown, but the leather is also brown. When we apply our test, therefore, the *and* does not sound right.

RULE 6. Never place a single comma between the subject and its verb, a verb and its object or complement, or a preposition and its object.

Wrong	Right
The environmental test program, was the cause of the third-quarter loss.	The environmental test program was the cause of the third-quarter loss.
The environmental test program produced, the third-quarter loss.	The environmental test program produced the third-quarter loss.
The environmental test program was the cause of, the third-quarter loss.	The environmental test program was the cause of the third-quarter loss.

Semicolon Rules

Like the comma, the semicolon is used to mark a logical discontinuity of thought. It is slightly stronger than the comma, but it is not quite as strong as the period. It is really needed in only two sentence constructions.

RULE 7. For clarity's sake, place a semicolon between items in a series containing internal commas.

> She has worked in Toronto, Ontario; Winnipeg, Manitoba; and Calgary, Alberta.

RULE 8. Place a semicolon between closely related independent clauses not joined by a coordinating conjunction.

This rule will come into play when you wish to use an adverb as a connector, or when you wish to omit a connecting word altogether, in order to create a stylistic effect. Some examples follow:

> The system completely failed the tests; therefore, requests for further funding have been denied.

> The system completely failed the tests; requests for further funding have been denied.

A causal relationship is shown between the two occurrences described in the examples; that is, further funding has been denied on the basis of the failure of the system to pass certain tests. In the first example, the word *therefore* clearly establishes the causal connection. In the second example, the presence of a semicolon without any connecting word creates a somewhat different stylistic effect.

Colon Rules

The colon is the mark of introduction. Its use is thus very restricted:

RULE 9. Place a colon after an independent clause or a sentence that introduces something.

Here are some examples of the correct use of the colon (including this very sentence):

> Operational tests will be conducted within three simulated environments: high altitude, low temperature, and high humidity.

> Before setting out, you will need several items of equipment: a waterproof tent, a week's supply of food, and a heavy caliber rifle.

The colon is also used following the salutation in business letters, within certain bibliographic styles, etc.

Hyphen Rules

Since compound words occur often in technical writing, writers of technical information should become familiar with the common ones in their particular fields. Words such as *livingroom* and *downtown* are typical examples. Such words have evolved from two words, through hyphenated words, to single words. Unfortunately, standard dictionaries often conflict with the common practices in certain fields and are, therefore, unreliable guides. A company style guide is usually a reliable reference for compound words.

RULE 10. Place a hyphen between words forming a unit-modifier.

Common examples of unit-modifiers occur when numbers and units of measure combine to act as single modifiers. Combinations such as *five-second delay* are typical. The hyphen is retained even if the number style requires a numeral: *5-second delay*. When the unit of measure is abbreviated, some styles retain the hyphen, as in *8-ft. studs*.

When the unit of measure is a symbol, many styles omit the hyphen, as in *10 kHz signal*.

Some word combinations are so familiar that writers may safely drop the hyphen when the combination is used as a modifier. *Stainless steel* is a good example of such a combination.

RULE 11. Place a hyphen between the prefix and the root word if awkward double vowels would otherwise result.

Re-entry is easily and quickly understood, whereas *reentry* takes an extra millisecond or two for the human brain to process. *Antiintegration* looks awkward, as does *antiaircraft* with its three successive vowels.

The problem of knowing when to hyphenate is exacerbated by the great number of common words, such as *cooperate and coincidence,* that are no longer hyphenated. Don't be afraid to consult a dictionary, but beware of those words that have a different meaning when hyphenated: *re-cover* and *re-cover,* for example.

Two Special Punctuation Problems

FUSED SENTENCES

A fused sentence is made up of two or more independent clauses without punctuation separating them.

> The sergeant called the roll he also issued the duty roster.

> The sergeant called the roll and he also issued the duty roster.

The first example requires a period or a semicolon to follow the word *roll*.

> The sergeant called the roll; he also issued the duty roster.

> The sergeant called the roll. He also issued the duty roster.

In the second example, a comma should have been placed after the word *roll,* or the word *he* should have been omitted.

> The sergeant called the roll, and he also issued the duty roster.

> The sergeant called the roll and also issued the duty roster.

Adding the comma changes the example from a fused sentence to a compound sentence. Omitting the word *he* changes it to a simple sentence with a compound verb.

COMMA SPLICES

A comma splice, also called a *comma fault*, occurs when a comma is made to do the work of a semicolon or a period. Comma splices occur most frequently because writers fail to realize that they have not used coordinating conjunctions. There are only five coordinating conjunctions: *and, but, or, nor,* and *for.* (Many authorities also now regard *yet* and *so* as coordinating conjunctions.) If one of these seven words is not present between independent clauses, the comma is an insufficient separator.

Conjunctive adverbs, such as *therefore, however, nevertheless, moreover, furthermore,* etc., can also be used to join independent clauses. Such words and word groups are often necessary to provide essential transitions. In this way, they act as conjunctions. However, since a conjunctive adverb's joining function is weaker than that of a coordinating conjunction, it must be preceded by at least a semicolon. This punctuation is necessary to signal the reader that the adverbial relationship is to the second clause only. Some examples follow:

Georgia saw Harold go into the theater; consequently, she knew he had lied to her.

The highway between Calgary and Banff is excellent; however, it can be hazardous in the winter months.

Mere commas coming before *consequently* in the first example and *however* in the second example would fail to signal to the reader that each of these connecting words modifies the second clause and not the first.

When these words are used as adverbs only, they are not set off by punctuation unless they are used parenthetically. The following examples illustrate these uses:

However great its artistic value may be, a violent television show is not necessarily suitable for viewing by all family members.

She was, however, very conservative with her money.

In the first example, *however* is an adverb modifying the adjective *great*. A comma is not placed after it because it must not be set apart from the word it modifies. In the second example, *however* is again used as an adverb, but it is parenthetic because it has also the transitional function of connecting the sentence to a previous thought.

Exercises

B-1. Punctuating Sentences

Place commas, semicolons, and colons in the following sentences to satisfy punctuation rules.

1. Every time students follow the steps in the writing process they learn something new.

2. You should work fast when writing a first draft.

3. The first three of the seven liberal arts which are called the *trivium* are grammar rhetoric and logic.

4. A misspelled word or a heavy style only distracts readers an ambiguous statement can result in disaster.

5. Poor word choices often give readers the wrong message.

B-2. *Using Hyphens*

Place hyphens in the following sentences to satisfy hyphen rules.

1. The technical writer who has learned to write only one kind of technical document while in school is limited by such one dimensional training.

2. The Bushmaster DM-3 has never been mass produced; the prototype is a one of a kind aircraft.

3. The site is defended by eight or more ground to air missile emplacements.

4. The test caused a ten centimeter long crack in the transmission housing.

5. The field engineering report contains a recommendation for a six hour burn in test.

Agreement

Subject-Verb Agreement

SIMPLE SUBJECT RECOGNITION

The first major contributor to subject-verb agreement errors is the failure of the writer to identify the simple subject. Words that come between the simple subject and verb can cause confusion:

(Wrong)	One of the competitors *were* found to be ineligible.
(Right)	One of the competitors *was* found to be ineligible.

Inverted sentences can also be deceiving:

(Wrong)	Just across the hall *is* the kitchen and the dining room.
(Right)	Just across the hall *are* the kitchen and the dining room.

SUBJECT–NUMBER RECOGNITION

The second reason, failing to know the subject's number, often has to do with collective words, such as *government, committee, team,* and *company.* All these words are singular. When you intend to refer to the individual members of a team or a committee or to the individual members of a legislature, you should avoid collective words. The following examples will help to clarify this principle:

(Wrong)	The team were given trophies.
(Right)	Each member of the team was given a trophy.
(Wrong)	The government were compelled to pass the new law.
(Right)	The government was compelled to pass the new law.

INDEFINITE PRONOUN SUBJECTS

The indefinite pronouns include *everybody, anybody, either,* and *neither.*
    Either and *neither* cause trouble occasionally because writers fail to realize that these words are singular when they occur without *or* or *nor.*

Either of them is acceptable.
Neither of them is acceptable.

When these two words appear with an *or* or a *nor,* they act as conjunctions between alternative subjects. In this usage, the number of the subject nearest the verb determines the verb form.

Neither George nor the other team members were aware of the oncoming train.
Neither the other team members nor George was aware of the oncoming train.

Although both of the foregoing examples are correct, most writers would agree that the first version sounds better and is preferable for that reason.

Pronoun-Antecedent Agreement

The same people who routinely make pronoun subjects and their verbs agree falter when pronoun gender enters the picture. This failure leads them to say or write sentences such as these:

Everybody who was impressed by the speaker was asked to raise *their* hand.

Anybody who wanted a second helping was asked to help *themselves.*

This practice, which has recently become commonplace, can be traced to a requirement for language that is inclusive of women. Many people now object to the versions shown in the following examples, in which the rule of agreement is observed:

> Everybody who was impressed by the speaker was asked to raise *his* hand.

> Anybody who wanted a second helping was asked to help *himself.*

In such cases, technical writers must find alternatives to traditional expression, rather than adhere simply to the agreement rule. The best course of action is to adhere strictly to the following general rule: *Never distract your readers*. Do not fall into the trap of the "he or she" formula, which can become a distraction. Also, avoid using "they" to refer to singular nouns, and shun the equally distracting alternation of masculine and feminine gender pronouns.

However, adherence to the "no-distractions" rule also commands the omission of inclusive masculine pronouns and possessive adjectives. Instead, impartial writers will recast distracting sentences using plural pronouns.

> Those who were impressed by the speaker were asked to raise their hands.

> People who wanted a second helping were asked to help themselves.

The no-distractions rule also commands writers to use *police officer* rather than *policeman, firefighter* rather than *fireman, letter carrier* rather than *postman,* etc.[1]

B-3. Correcting Subject-Verb Agreement Errors

Make all the subjects in the following sentences agree with their verbs.

1. A full range of engineering and test facilities are available.

2. The water supply for Washburn County residents come from a reservoir.

3. As shown in the table, all the units was installed in December.

---

[1] For additional examples of bias-free terminology, you may wish to consult Rosalie Maggio, *The Bias-Free Word Finder, a dictionary of nondiscriminaatory language.* Boston: Beacon Press, 1991.

4. The committee were impressed by the demonstration.

5. One of the committee members were taken ill during the demonstration.

B-4. *Ferreting Out Pronoun-Antecedent Errors and Other Distracting Words*

Rewrite the following sentences to remove distracting words.

1. The firemen were unable to contain the blaze.

2. Each participant was asked to record his or her results on the forms provided.

3. Each team member has a computer at their own desk.

4. The technician found the error and recorded his finding in his logbook.

5. The board of directors were not involved in the decision.

Words That Mean What You Are Trying To Say

Malaprops (or Malapropisms)

Words signify reality. When we use a word with which we are only vaguely familiar, we sometimes give the reader a signal that we didn't intend. We refer to such errors as malaprops. Unintentional malaprops betray an undisciplined mind at work.

Adverse AND Averse

In the statement, "He was not adverse to war if the cause was just," the word *adverse* (acting against) is a malaprop, mistaken for *averse* (having an aversion to). A correct use of *adverse* would be, "She had an adverse reaction to the drug."

Comprise AND Compose

A thing cannot be "comprised of" something else. The word *comprise* is synonymous with the word *include*. The whole comprises the parts; the parts compose the whole.

(Wrong)	Water is *comprised of* hydrogen and oxygen.
(Right)	Water is *composed of* hydrogen and oxygen.
(Right)	Water *comprises* hydrogen and oxygen.

Extraneous Words

Some words and phrases contribute little or no meaning to the message. Some common ones are listed here:

basically

it has been brought to my attention that

it seems as if

it is my opinion that

the purpose of this letter is to

Redundant Words

Redundancy means saying the same thing twice. The italicized portions of the phrases shown below are redundant and should be omitted.

purple *in color*

return *back*

I know *for a fact* that

each *and every*

exactly identical

true facts

Wordy Expressions

The phrases in the left-hand column below are not intrinsically bad. Unskilled writers can nearly always improve their writing by replacing the phrases shown in the left-hand column with the corresponding single words in the right-hand column.

Change ...	To
due to the fact that	because (or since)
notwithstanding the fact that	although
at the present time	now
at that point in time	at that time (or then)
in many cases	often
in most instances	usually
in the event that	if
with regard to	about
in the amount of	for
during the time that	while

Many phrases can be reduced by substituting a specific verb:

Change ...	To
have a meeting	meet
make a selection	select
advance an argument	argue
make a concerted effort	try
be in agreement with	agree
be of the opinion that	think

B-5. Identifying Malaprops

Substitute the right word for any malaprop you discover in the following sentences. Rewrite the sentence if you have to.

1. If the mine doesn't reopen soon, the company will flounder.

2. These hazards are comprised of a variety of conditions.

3. Hopefully, these conditions will be corrected before the mine reopens.

B-6. *Getting Rid of Extraneous Words and Replacing Wordy Expressions*

Rewrite the following sentences to remove wordy expressions.

1. I am writing this letter to tell you that we no longer stock Part No. 488119H.

2. This report is on the topic of the AN/ASQ-21 Personnel Detection System.

3. The system operated properly notwithstanding the fact that the alarm was a false one.

4. The launch will go ahead as planned in view of the fact that the weather has cleared.

5. The employees are of the opinion that the committee has put its own interests ahead of those of the membership.

Sentences That Say What You Want Them to Say

The sentence, the basic pattern of verbal thought, is the foundation upon which all English prose is built.

Each word in a sentence has a distinct job to do. Words used unnecessarily and inappropriately are more than errors in grammatical convention: they are errors in logic. As such, they not only prevent communication from taking place, they sometimes fool listeners and readers into thinking that communication has taken place when it hasn't.

Even relatively minor grammatical errors can frustrate and confuse readers. Lapses in logic distract or frustrate readers. Serious mistakes can destroy meaning.

When your ear tells you something isn't quite right about a sentence, you may be reacting to an unfamiliar style and not to incorrect grammar. In order to help you distinguish grammar from style, each of the categories of sentence errors discussed herein is followed by examples.

Expletive Constructions

In some sentences beginning with the words *it* or *there*, the first word carries no meaning of its own, but rather substitutes for an idea expressed later. Such constructions can be overused. In the following examples, the expletive constructions are more of a barrier than an aid to clarity.

It is probably a mistake to attribute the accident to engine failure.

The pronoun *it* is a substitute for the infinitive phrase *to attribute.*

There was an anomaly in the test results.

The adverb *there* is a substitute for the adverb phrase *in the test results.*

Such sentences can often be improved by restructuring. Find the real subject and put it at the beginning of the sentence; then find the real action of the sentence and express it as a verb.

Change ...	To		
It is probably a mistake to attribute the accident to engine failure.	To attribute the accident to engine failure is probably a mistake.	*or*	Attributing the accident to engine failure is probably a mistake.
There was an anomaly in the test results.	An anomaly appeared in the test results.		

Faulty Predication

Faulty predication is produced when the writer fails to logically match the subject and predicate.

THE VERB *To Be*

Often, a form of the verb *to be* leads to a faulty predication.
For example:

The cost of obtaining an education today is expensive.

This example says the *cost* ... is *expensive.* Such a statement is illogical. It is a result of mixing two methods of making the statement:

The cost of obtaining an education today is very great.

Obtaining an education today is expensive.

Both of these sentences are correct, but they cannot be logically combined.

TRANSITIVE VERBS

Ordinary transitive verbs can also be improperly matched.
For example:

The question of whether people should wear seat belts infringes our freedom.

This sentence says that the *question infringes*, but its intention is probably to say that a law requiring people to wear seat belts would infringe.

Where AND *When*

Other common faulty predications result from the coupling of a noun, pronoun, or gerund with a clause beginning with where or when.

| (Wrong) | Plagiarism is where one steals a writer's work. |
| (Right) | Plagiarism is the theft of a writer's work. |

| (Wrong) | Happiness is when you win the game. |
| (Right) | Happiness is the feeling you get when you win the game. |

Reason WITH *Because*

Another common faulty predication comes about when *the reason is* is coupled with *because*.

| (Wrong) | The reason for her missing Saturday's game was because she had injured her shoulder. |
| (Right) | The reason for her missing Saturday's game was that she had injured her shoulder. |

Pronoun Reference

A pronoun is a word that takes the place of a noun. An error in pronoun reference occurs when one uses a pronoun without giving a clear idea of what noun it refers to (its antecedent). Here is an example of an error in pronoun reference:

Donald told Bill that he had made a mistake in his estimate.

Did Donald make a mistake, or did Bill make a mistake? *He* could refer to either one. Here is an example that might be found in a technical document:

The machine welds the steel tubing together at a certain specified point in response to instructions from the computer, which has been programmed to indicate the selected junction to the automatic system. This permits the shipping of completed subassemblies.

In the foregoing example, the word *this* has no antecedent. Did the writer mean *this machine, this weld, this computer, this junction,* or *this automatic system*? The writer's intention might have been to have *this* refer to the entire process. Such writing is careless and potentially confusing. Therefore, the writer should have written the sentence as follows:

This process permits the shipping of completed subassemblies.

In this version, the word *this* is a demonstrative adjective and not a pronoun.

The relative pronoun *which* is also sometimes made to refer to something other than a single-word antecedent.

We had plenty of snow during March and April, thereby averting drought this year, *which* was lucky for the farmers.

The writer probably intended the word *which* in this example to stand for the entire idea that drought was averted. Sentence logic suggests that *which* refers to the noun that precedes it. A reader who follows this logic will interpret the clause beginning with *which* to mean that the *year* was lucky for the farmers. At best, even if readers understand the sentence, the writer has made such understanding more difficult by failing to make the pronoun stand for a single noun.

Some writers compound the error by using *which* as a connector to loosely string thoughts together.

My car broke down, which caused me to miss my plane, which spoiled my whole weekend.

Remember that *which* is a pronoun; whenever it is used to introduce an adjective clause, it must represent a noun, and that noun must be instantly apparent to the reader. The sentence above could be improved as shown below:

I missed my plane because my car broke down, and my whole weekend was spoiled.

It WITHOUT AN ANTECEDENT

Careless writers sometimes use *it* to refer to an adverb phrase.

In the newspaper, it says that the Japanese Yen is gaining value.

Structured this way, the sentence means *in the newspaper, the newspaper says that the Japanese Yen is gaining value.* The sentence should be written as follows:

The newspaper says that the Japanese Yen is gaining value.

or

An article in the newspaper says that the Japanese Yen is gaining value.

VAGUE REFERENCES USING *They*

The pronoun *they* should never be used to refer to unspecified persons. This error occurs most often in sentences such as the following:

In Toronto, they put salt on the roads to melt the ice during the winter.

If you have a particular subject in mind, name it:

In Toronto, road crews put salt on the roads to melt the ice during the winter.

If you do not want to specify a subject, cast the sentence in a different way:

In Toronto, salt is used for melting ice on the roads during the winter.

CARELESS USE OF *You*

An objective viewpoint is often desirable, but the reluctance to use *I* sometimes leads to inappropriate uses of *you.*

As the tornado approached, you could see parts of trees and buildings being sucked up into the vortex.

The pronoun *you* should be used only if the writer intends the reader to be involved in the sentence. Such use is appropriate for giving advice or instructions and for asking questions, but not for referring to persons other than the reader. Here are some better ways of writing the above sentence:

As the tornado approached, I saw parts of trees and buildings being sucked up into the vortex.

or

As the tornado approached, people near it could see parts of trees and buildings being sucked up into the vortex.

Only one rule applies to pronoun reference: *Never use a pronoun without a clear antecedent.*

Pronoun Case

Many pronouns have one form that is used when the pronoun represents a subject or a subject complement and a different form when the pronoun represents an object of a verb or preposition. For example, we would use the pronoun *I* as the subject in the sentence *I lost my dog*, but we would use the pronoun *me* (representing the same person) in the sentence *My dog bit me*.

COMPOUND SUBJECTS AND OBJECTS

When a pronoun is joined to a noun by a coordinating conjunction, the pronoun should be in the same case as the noun. Try reading the sentence without the noun; if the pronoun sounds right by itself, it is probably in the correct case.

The dog followed Billy and I to the park.

Drop the noun, and read the sentence.

The dog followed ... I to the park.

The sentence sounds wrong because the objective pronoun *me* should have been used.

The dog followed Billy and me to the park.

The following sentences show pronouns used in the correct case.

It was I who telephoned. (subject complement)

Emiline persuaded Wally and me to play golf with her. (object [followed by an object complement])

That boat belongs to her and her sister. (object of a preposition)

PRONOUNS IN APPOSITION TO NOUNS

When a pronoun is placed beside a noun as an appositive of that noun, the pronoun should be in the same case as the noun. You can check the case by ignoring the noun and reading the sentence with just the pronoun. The following sentences illustrate correct uses:

Examples with pronouns and nouns

We students work very hard.

It is *we* the *students* who will suffer most from this new policy.

The bursar told *us students* that we would have to work harder.

The party was for *us students*.

Examples with pronouns only

We work very hard.

It is *we* who will suffer most from this new policy.

The bursar told *us* that we would have to work harder.

The party was for *us*.

Who OR *Whom*

These two words cause much confusion. The confusion is really unwarranted, however. As a test, simply substitute *he* or *him*. *He* corresponds with *who*, and *him* corresponds with *whom*.

The words *who* and *whom* are used in questions, in adjective clauses, and in noun clauses. Use *who* to represent a subject or a subject complement; use *whom* to represent an object of a verb or a preposition.

As interrogative pronouns in questions

Who took my toothbrush?

Whom did you see at the store?

With whom are you going to Europe?

As relative pronouns in adjective clauses

The lady who owns that dog is a local politician.

Myriam, who I hoped would win, came in second. (The verb *would win* needs a subject.)

My dentist, whom I see twice a year, does excellent work.

As introductory words in noun clauses

I know who took your toothbrush.

I do not know who that man beside your brother is.

Nobody knows whom Bernice will choose to be Assistant Manager.

Lana did not say whom the gift was for.

Whoever OR Whomever

Use *whoever* to represent a subject or a subject complement; use *whomever* to represent an object or an object of a preposition. The following sentences illustrate correct uses.

Subject in a noun clause

Whoever took my toothbrush is in big trouble. (subject complement in a noun clause)

That man, whoever he is, has been standing there for two hours.

I saved this space for whoever wants it. (Note in this example that the whole noun clause [not just the word *whoever*] acts as the object of the preposition *for.*)

Object in a noun clause

The old man will give the trophy to whomever he likes best. (Note in this construction that the whole noun clause acts as the object of the preposition *to.* In the noun clause, *he* is the subject, and *whomever* is the object.)

Modifier Placement Errors

English nouns and modifiers stay the same no matter what purpose they serve. This feature of English gives its users a versatility and freedom of expression found in no other language, but it also makes the positioning of words very important. Note how the meaning changes in the following examples, all of which have the same words with the word *only* moved around:

adjective

Only I wish George had been the winner. (I alone wish George had been the winner.)

<u>adverb</u>

I only wish George had been the winner. (My only wish is that George had been the winner.)

<u>adjective</u>

I wish only George had been the winner. (I wish that George did not have to share his win with others.)

<u>adjective</u>

I wish George had been only the winner. (I wish that George had been a winner and nothing else.)

<u>adjective</u>

I wish George had been the only winner. (I wish that George had been the only one to win.)

Many different meanings are produced merely by changing the position of one modifier in a sentence. Be careful to position modifiers next to the words they modify. A one-word modifier should usually precede the word it modifies, and a multiple-word modifier (phrase or clause) should usually follow the word it modifies.

A dangling modifier is a modifying word, phrase, or clause that has nothing to modify. The following example illustrates such an error.

Considering the complex nature of modern government, it is no wonder that corruption is widespread.

Who is considering? The phrase dangles because it cannot modify the grammatical subject of the sentence. An introductory phrase must modify the grammatical subject. Several ways to correct the example are available.

A grammatical subject can be supplied:

Considering the complex nature of modern government, one shouldn't wonder how corruption can be widespread.

The introductory phrase can be changed to a dependent clause:

When people consider the complex nature of modern government, they will not wonder how corruption can be widespread.

The entire sentence can be recast to convey the thought in another way:

The complex nature of modern government makes corruption inevitable.

Parallel Structure

Parallel structure matches grammar to logic. Two things that are equal in importance or rank should have the same grammatical form. Two things having an unequal relationship should have that inequality reflected in the grammatical structure.

Words and expressions connected by *and, but,* and *or* lead the reader to expect logical equality. Therefore, such words and expressions must be made grammatically equal. An example follows:

> My hobbies are to fish in the summer and cross-country skiing in the winter.

In this example, the infinitive phrase *to fish* is coupled with the gerund *skiing,* thereby mismatching the logic and grammar of the sentence. The mismatch may be solved in two ways:

> My hobbies are fishing in the summer and cross-country skiing in the winter.

> I like to fish in the summer and to cross-country ski in the winter.

Here are more examples of faulty parallel structure:

(Wrong)
    Applicants for the program were required to be good
    <u>noun</u>   <u>adjective</u>
    scholars and honest.

                                                 <u>adjective</u>

(Right)
    Applicants for the program were required to be scholarly
     <u>adjective</u>
    and honest.

                     <u>adjective</u>

(Wrong)
    They appeared unfriendly and as though they
    <u>clause</u>
    were apprehensive.

                   <u>adjective</u>   <u>adjective</u>

(Right)
    They appeared unfriendly and apprehensive.

                   <u>gerund phrase</u>   <u>clause</u>

(Wrong)
    The budget allowed for painting the walls, or the ceilings
    could be replastered.

	gerund phrase gerund phrase
(Right)	The budget allowed for painting the walls or replastering the ceilings.

	prepositional phrase dependent clause
(Wrong)	We were hoping for rain or that the heat would let up at least.

	dependent clause dependent clause
(Right)	We were hoping that it would rain or that the heat would let up at least.

Consider the following example:

The band leader was late and the chorus girls angry.

The non-parallel nature of the example comes about because the singular word *was* cannot be applied to the plural phrase *chorus girls.* Also, the error in parallel structure leads to a second error, the omission of the comma that should logically separate the two thoughts. A correct version of this sentence is as follows:

The band leader was late, and the chorus girls were angry.

A series must also adhere to parallel structure because the items therein contained are always logically parallel.

My hobbies are fishing, hunting, and surfing. (all gerunds)

I like to fish, to hunt, and to surf. (all infinitives)

Mary has beauty, charm, and grace. (all nouns)

Mary is beautiful, charming, and graceful. (all adjectives)

A list is also always parallel by its nature. The following example, a typical list, is made up of noun phrases:

Three characteristics are required of an applicant for the police department:

1. Good physical condition

2. A stable mental history

3. An eagerness to serve the public

Faulty Coordination/Subordination

Faulty coordination/subordination places logically unequal elements in logically equal structures.

The following example illustrates this idea:

Snow fell for three days, and the RCMP closed the highway.

Although no grammatical fault is to be found in the example, it may not say exactly what its author intended. Subordinating one of the two independent clauses would give us the following complex sentence in place of the compound sentence:

The RCMP closed the highway because snow had fallen for three days.

Or:

The RCMP closed the highway because of a three-day snowfall.

Or, even more simply:

A three-day snowfall forced the RCMP to close the highway.

Shifts in Voice and Mood

Active and passive main clauses should not be combined in the same sentence. Here is an example of such a combination:

Laurie likes pizza, but spaghetti is preferred by Florence.

The foregoing is a compound sentence, containing both the active verb *likes,* and the passive verb *is preferred.* Sentences such as this one require the reader to mentally shift gears. It should read as follows:

Laurie likes pizza, but Florence prefers spaghetti.

Active and passive structures should not be combined in complex sentences or in consecutive sentences if a subsequent verb is made to refer back to the verb of the first clause or sentence. Forms of the verb *to do* are often found in such mixed voice constructions. Examples follow:

(Wrong)	Since a way of acknowledging donations had to be found, the provincial government has done so.
(Right)	Since someone had to find a way to acknowledge donations, the provincial government has done so.

(Wrong)	The book can be read in two hours. If you do so, you won't be sorry.
(Right)	The book can be read in two hours. If you read it, you won't be sorry.

Instructional steps are normally written in second-person, imperative mood. They must not contain a mixture of person and mood, as does the following example:

1. Remove the hubcap from the wheel by prying it off with the flat end of the lug wrench.
2. The car should be jacked up at this point.
3. Remove the wheel.
4. The spare wheel should be placed on the hub and fastened loosely with one lug nut.

Steps 2 and 4 conflict with the others because they are written in the third-person, subjunctive mood.

Faulty Comparisons

OMITTED WORDS

Faulty comparisons often result from the omission of words. Consider the following sentence:

Bart is as good at the quarterback position if not better than Ralph is.

The second *as* has been omitted. When we compare two things, we must say *as good as,* not *as good than.* The sentence could be corrected as follows:

Bart is as good at the quarterback position as Ralph is, if not better.

Sometimes we omit an essential clause element, as shown in the following example:

Greta taught me more about mathematics than Phyllis.

Did Greta teach me more than she taught Phyllis? Or did Greta teach me more than Phyllis taught me?

Greta taught me more about mathematics than she taught Phyllis.

Greta taught me more about mathematics than Phyllis did.

Many comparisons are faulty because the word *than* is omitted, leading to the mention of only one item, as in the following example:

Ben's cooking is much better.

Is Ben's cooking better than some other person's cooking? Is it better than it used to be?

Another common faulty comparison comes about when a person or thing in a group or set is declared to be more or most without an accompanying word to indicate the person or thing's membership in the group. The following example illustrates this error.

(Wrong)   Sally is more daring than anyone on the soccer team. (The inference to be drawn is that Sally is not on the soccer team.)

(Right)   Sally is more daring than anyone else on the soccer team. (The addition of *else* makes it clear that Sally is a member of the soccer team.)

Like AS A CONJUNCTION

Another faulty comparison results from using like as a conjunction, as in the following advertising slogan:

Winston tastes good like a cigarette should.

The word *as* must be used in place of *like* in such constructions if sentence balance is to be maintained. *Like* links substantives, not whole clauses.

B-7. *Identifying and Eliminating Expletive Constructions*

Rewrite the following sentences to get rid of expletive constructions.

1. There were too many assemblies rejected during the calibration stage.

2. It is wrong to attribute the rejections to poor design.

3. There is only one way to fix the problem, and that is to rewrite the manufacturing process procedures.

4. There is another reason for rewriting the process procedures.

5. It shouldn't come as a surprise to anyone that the procedures also waste the readers' valuable time with far too many expletive constructions.

B-8. *Identifying and Eliminating Faulty Predications*

Rewrite the following sentences to get rid of faulty predications.

1. The expense of rewriting the manuals will be quite costly.

2. Changes in the manual will be different from the old manual.

3. The only failures where engineering was the cause were in the subassembly stage.

4. The threat of failure in this device is dangerous.

5. A good product is when it works perfectly.

B-9. *Identifying and Eliminating Pronoun Reference Errors*

Rewrite the following sentences to correct pronoun reference errors.

1. This means that workers must know all the operations, not just their own.

2. The statement is an example of the wrong information in the manual, which could cost some employees their jobs.

3. One part of the manual is particularly misunderstood, which is the final assembly.

4 This must be corrected through the issuance of a new manual as soon as possible.

5. This is the only way to clear up misunderstandings.

B-10. *Choosing the Correct Pronoun Case*

Rewrite the following sentences to correct pronoun case errors. Some of the sentences may not contain errors.

1. Give the technical writing group leader's job to whomever wants it.

2. The supervisor told us technicians to get the job done fast.

3. Does anyone know whom Madeleine chose to be her new assistant?

4. Gregory's request for a salary review is between he and the chief engineer.

5. Us model shop employees are the last to be told about changes in policy.

B-11. *Identifying and Eliminating Modifier Errors*

Rewrite the following sentences to eliminate dangling and misplaced modifiers. Some of the sentences may not contain errors.

1. Wondering what to do next, the fault corrected itself.

2. Hoping that the software was bug-free, the project went ahead.

3. Violating a safety rule once was considered grounds for dismissal.

4. Exhausted after the long flight, a hot shower and soft bed were the only things the pilot could think of.

5. Smelling of liquor, the police officer arrested the motorist.

B-12. *Identifying and Eliminating Non-Parallel Structure*

Rewrite the following sentences to eliminate non-parallel structure. Some of the sentences may not contain errors.

1. This manual contains removal, disassembly, reassembly, reinstallation, and testing procedures.

2. Evaluate the test results in accordance with specification paragraph 11.9 and the criteria given in specification paragraph 12.2.

3. The support facilities available for this project include a complete CAD/CAM setup, an environmental test laboratory, and the model shop is available.

4. The sensors feed into a data conversion unit, which will receive, process, encode, and provide the decoding step.

5. The functions of the unit are gathering, processing, and to transmit data.

B-13. *Using Coordination and Subordination*

Combine the sentences given below, coordinating and subordinating grammatical structures to match the logic of the thoughts expressed. Use coordinating and subordinating conjunctions or other transitional words and phrases as needed.

> Direct-mail marketing is an art. Most people don't know about the techniques direct-mail copywriters use. Some direct-mail techniques are related to graphic design. Graphic design is related to the way reading has condi-

tioned us. A reader's eye moves from left to right across a page. A reader's eye moves from the top of a page to the bottom. The direct-marketing copywriter knows how our eyes move. The direct-marketing copywriter assists the reader's eye. Strategically placed graphic features assist the reader's eyes. Graphic design helps convey the message.

B-14. *Identifying and Eliminating Voice and Mood Shifts*

Change the following sentences to get rid of unnecessary shifts. Some sentences may not contain errors. (Please notice that voice and mood shifts are related to non-parallel structure, especially mood shifts in procedural steps.)

1. Some people think that using the passive voice is always wrong, but the passive voice is appreciated by those who know when to use it.

2. Rebel against your grammar checker, and you should rebel today.

3. Sometimes, writers don't know who performed an action, and the active voice cannot be used.

4. Writing in which too many passive constructions have been used by the writer makes the reader's job more difficult than it should be.

5. The active-voice sentence can be read easily because it follows the natural sequence of an event.

B-15. *Identifying and Eliminating Faulty Comparisons*

Change the following sentences to get rid of faulty comparisons. Some sentences may not contain errors.

1. An isotope-fueled thermoelectric generator would be more harmful to the environment.

2. Fuel cells do not represent an environmental threat any more than solar power.

3. A full-time gasoline-powered system would be even worse.

4. Our design poses less threat to the environment than any other combination.

5. Our design is environmentally friendly, like a good design should be.

Spelling

Canadian Dictionaries and Spelling Practices

Canadian English dictionaries do not offer uniform spelling choices. Spelling practices also vary from one provincial education department to another, from school board to school board, and from newspaper to newspaper. The real problem is that Canada has no unabridged dictionary of its own.

Canadian manufacturers and service industries that do business in the United States adopt Webster spellings in technical documents tailored for American customers. And the proliferation of computer spell checkers has tended to move Canadian spelling practices in that same direction. Even recent editions of the Oxford Dictionary show a movement toward Webster spellings.

The spellings in this book conform to the *Canadian Edition of the New Lexicon Webster's Encyclopedic Dictionary of the English Language* and, coincidentally, to the computer spell-check program used in its composition. As a technical writer, you will want to adhere to the spelling practices common to your industry and to your particular firm.

Using a Dictionary

Although dictionaries differ slightly, all of them include a good deal of information within each entry. For example, although some entries may show two spellings, the first spelling is the preferred spelling; the second is included to indicate that another spelling may have been common at one time.

Correct pronunciations are usually given next, again with the preferred pronunciation coming first.

Next, the part of speech (noun, verb, adjective, etc.) is given. When a word can be used as more than one part of speech, separate definitions are given for each.

Etymology, or word origin, is usually given next. Such information is useful not only in providing an understanding of the original meanings of words but in giving clues to the relationships of words to each other. The word *tenet*, for example, comes to English directly from Latin, in which it means *he holds*. In English, the word is a noun meaning a dogma, principle, belief, or doctrine *held* to be true. Many other words have come to us from the same origin.

tenant	One who temporarily holds property.
tenacious	An adjective applied to anything that holds fast
lieu**ten**ant	One who holds power in **lieu** of (from French, in place of) a superior

Dictionaries also give definitions specific to given fields, as illustrated in the following definitions of *moment:*

1. A minute portion of time; an instant.
2. Importance, as in influence or effect.
3. A definite period or point; specif., Chiefly Philos., a stage in development.
4. An essential or constituent element.
5. Mech. Tendency, or measure of tendency, to produce motion.

Finally, dictionary entries give a synonym for the word.

Prefixes

Prefixes are letters attached to the beginnings of words for the purpose of changing their meanings in some way. Some common prefixes are *dis, mis, il, un, in,* and *im.* The general rule governing prefixes is that they do not affect the spelling of the words they precede.

Certain prefixes can precede only adjectives, adverbs, and verbs. Others can precede only nouns. The following list contains correctly spelled words beginning with prefixes.

dis+please	= displease	un+necessary	= unnecessary
mis+print	= misprint	in+satiable	= insatiable
il+logical	= illogical	im+mature	= immature

Suffixes and the Final Consonant — To Double or not to Double

Suffixes, like prefixes, are added to words to change their meanings in some way. The suffix is attached to the end of the word; sometimes the original word's spelling must be changed before the suffix can be added.

The suffixes *ly, ness, al, ment, less, ful, est,* and *er* can often be added to a word without changing its spelling. However, if the word ends with a *y*, the *y* is sometimes dropped or changed to an *i* before the suffix is added, as in the words *happily* and *mightily.*

Some suffixes, such as *ing* and *ed,* do require a spelling change in the root word before the suffix can be added. The rule is as follows:

A word ending with a single consonant preceded by a single vowel requires the doubling of its final consonant if the accent falls on the final syllable. Of course, if a word ends with two consonants, doubling is not required.

Note how each word in the following list obeys this rule.

occurring (accent on second syllable)

traveling (accent on first syllable)

dismissing (double consonant)

Note: The matter of *able* and *ible* is quite complex. The best course of action is to use a dictionary to solve all *able* and *ible* questions.

Ie and Ei words

The spelling *ei* is used after the letter *c* or when the syllable in question is pronounced as a long *a.* The rhyme that helps many people to remember the rule is as follows:

I **before** ***e*** **except after** ***c,*** **or when pronounced** ***a*** **as in neighbor and weigh.**

A few examples follow:

achieve	fiend	neighbor	shield
beige	fierce	niece	sleigh
conceited	freight	receive	vein
conceive	heinous	relieve	weigh

Exceptions to the rule must be memorized. They are as follows:

either	foreign	leisure	seize	species
financier	height	neither	sleight	weird

Spelling and the Apostrophe

The apostrophe has two uses: the formation of possessives and contractions.

The following single rule is sufficient to guide you in using the apostrophe to form possessives:

> Place an apostrophe and an *s* *('s)* after any noun you wish to put in possessive form, and then pronounce the word thus formed. If the pronunciation is incorrect, omit the final *s*.

Some examples are given in the following list:

Jones + 's = Jones's   boys + 's = boys's = boys'

men + 's = men's   Charles + 's = Charles's

Moses + 's = Moses's = Moses'

If you always pronounce such words correctly, you will find the foregoing rule to be foolproof. Otherwise, you may wish to refer to the following, somewhat more complex rule:

> Form the possessive of plural nouns not ending in s and all singular nouns by adding *'s*. If a singular possessive thus formed produces a triple *s* sound or a *z,s,s* sound, omit the final *s*. Form the possessive of plural nouns ending in s by adding an apostrophe only.

The formation of contractions is irregular and is related to the verb that forms part of the contraction. The words *don't, doesn't, it's, she's, they're*, etc., are all formed by placing the apostrophe in the position of one missing vowel. In the contractions *I've, we've, we'd, you'd, they'd*, etc., the apostrophe takes the place of at least one consonant and at least one vowel. *Won't* is in a class by itself, standing for *will not*.

Troublesome Words

The following list contains words that are often misspelled. Use a dictionary to learn the meanings of any of the words with which you are unfamiliar, paying special attention to homophones.

absence	analyze	consul
accessible	announce	contractual
accommodate	appraise	council
accredit	apprise	counsel
accumulate	ascent	
achieve	ascetic	decent
achievement	assent	deductible
acknowledgment		defense
address	cite	dependent
aesthetic	commitment	descent
affect	comparatively	diminutive
aforementioned	conceit	dissatisfy
aggravate	conscientious	dissent
alleviate	conscious	dissimilar

continued →

effect	its	prevalent
embarrass		principal
eminent	local	principle
environment	locale	procedure
exaggerate		proceed
except	mathematics	procession
	meter	prosecute
factual		
farther	necessary	reminiscent
foreword		remittance
formally	occasion	rhythm
formerly	occurred	
forty	omission	schedule
forward		seize
fourth	passed	separate
freight	past	site
further	persecute	solicit
	persevere	stationary
harass	personal	stationery
harassment	personnel	superintendent
	phenomenon	supersede
immanent	possess	tangible
imminent	practical	technician
incentive	practice	transition
independent	precede	vacuum
it's	precession	
	prerogative	

B-16. Using a Dictionary

1. Using a good dictionary, look up the words in the following list. Identify the preferred and alternate spellings, preferred and alternate pronunciations, etymology, parts of speech, usage level, synonym, common definition, and specialized definitions (if any) of each one.

2. Make up sentences using each word in the list. Write a separate sentence for each part of speech identified and for each specialized definition.

access	cite	effect	personal
accommodation	conceit	engineer	personnel
accredit	contractual		principal
aesthetic	council	farther	principle
affect	counsel	foreword	
appraise		formally	site
apprise	descent	further	stationary
ascent		harass	

B-17. Spelling

Identify the misspelled words in the list below, and spell them correctly in the space provided. Place a "C" in the space next to any word that is spelled correctly.

accessable	_____	freight	_____
accomodation	_____	further	_____
accredit	_____	harrass	_____
accummulate	_____	occured	_____
acetic	_____	occurence	_____
acknowledgment	_____	perogative	_____
appraise	_____	perservere	_____
apprise	_____	personell	_____
ascent	_____	practicle	_____
ascetic	_____	precession	_____
assent	_____	principal	_____
cite	_____	principle	_____
comparitively	_____	procede	_____
concientious	_____	procedure	_____
conciet	_____	reminiscent	_____
consul	_____	remittence	_____
counsel	_____	schedule	_____
decent	_____	sieze	_____
deductible	_____	site	_____
dissent	_____	stationary	_____
embarass	_____	stationery	_____
enviroment	_____	supersede	_____
farther	_____	tangible	_____
formally	_____	technition	_____
formerly	_____	transition	_____
fourth	_____		

Answers to Part B Exercises

B-1. *Punctuating Sentences*

(p. 353)

1. Every time students follow the steps in the writing process, they learn something new.

2. You should work fast when writing a first draft.

3. The first three of the seven liberal arts, which are called the *trivium*, are grammar, rhetoric, and logic.

4. A misspelled word or a heavy style only distracts readers; an ambiguous statement can result in disaster.

5. Poor word choices often give readers the wrong message.

B-2. *Using Hyphens*

(p. 354)

1. The technical writer who has learned to write only one kind of technical document while in school is limited by such one-dimensional training.

2. The Bushmaster DM-3 has never been mass-produced; the prototype is a one-of-a-kind aircraft.

3. The site is defended by eight or more ground-to-air missile emplacements.

5. The test caused a ten-centimeter-long crack in the transmission housing.

6. The field engineering report contains a recommendation for a six-hour burn-in test.

B-3. *Correcting Subject-Verb Agreement Errors*

(p. 356)

1. A full range of engineering and test facilities <u>is</u> available.

2. The water supply for Washburn County residents <u>comes</u> from a reservoir.

3. As shown in the table, all the units <u>were</u> installed in December.

4. The committee <u>was</u> impressed by the demonstration.

5. One of the committee members <u>was</u> taken ill during the demonstration.

B-4. Ferreting Out Pronoun-Antecedent Errors and Other Distracting Words

(p. 357)

1. The firefighters were unable to contain the blaze.

2. The participants were asked to record their results on the forms provided.

3. All team members have computers at their own desks.

4. The technician found the error and recorded the finding in the logbook.

5. The board of directors was not involved in the decision.

B-5. Identifying Malaprops

(p. 359)

1. If the mine doesn't reopen soon, the company will <u>founder</u>.

2. These hazards are <u>composed</u> of a variety of conditions.

3. <u>It is to be hoped that</u> these conditions will be corrected before the mine reopens.

B-6. Getting Rid of Extraneous Words and Replacing Wordy Expressions

(p. 360)

Other solutions are possible.

1. We no longer stock Part No. 488119H.

2. This report is on the AN/ASQ-21 Personnel Detection System.

3. The system operated properly although the alarm was a false one.

4. The launch will go ahead as planned since the weather has cleared.

5. The employees believe the committee has put its own interests ahead of those of the membership.

B-7. Eliminating Expletive Constructions

(p. 373)

Other solutions are possible.

1. Too many assemblies were rejected during the calibration stage.

2. Attributing the rejections to poor design is wrong.

3. The one way to fix the problem is to rewrite the manufacturing process procedures.

4. There is another reason for rewriting the process procedures. [The expletive construction in this sentence is an efficient way of making the statement.]

5. The procedures also waste the readers' valuable time with far too many expletive constructions.

B-8. Identifying and Eliminating Faulty Predications

(p. 374)

Other solutions are possible.

1. The expense of rewriting the manuals will be quite high.

2. The new manual will be different from the old manual.

3. The only failures caused by engineering occurred in the subassembly stage.

4. Failure of this device can cause a dangerous situation.

5. A good product is one that works perfectly.

B-9. Identifying and Eliminating Pronoun Reference Errors

(p. 374)

Other solutions are possible.

1. This statement means that workers must know all the operations, not just their own.

2. The statement, which could cost some employees their jobs, is an example of the wrong information contained in the manual.

3. The part of the manual that discusses the final assembly is particularly misunderstood.

4. & 5. Issuing a new manual as soon as possible is the only way to clear up misunderstandings.

B-10. Choosing the Correct Pronoun Case

(p. 374)

1. Give the technical writing group leader's job to <u>whoever</u> wants it.

2. The supervisor told us technicians to get the job done fast. (No changes)

3. Does anyone know whom Madeleine chose to be her new assistant? (No changes)

4. Gregory's request for a salary review is between <u>him</u> and the chief engineer.

5. <u>We</u> model shop employees are the last to be told about changes in policy.

B-11. Identifying and Eliminating Modifier Errors

(p. 375)

Other solutions are possible.

1. While the technician was wondering what to do next, the fault corrected itself.

2. Hoping that the software was bug-free, we went ahead with the project.

3. Violating a safety rule was once considered grounds for dismissal.

4. Exhausted after the long flight, the pilot could think only of a hot shower and soft bed.

5. The police officer arrested the motorist, who smelled of liquor.

B-12. *Identifying and Eliminating Non-Parallel Structures*

(p. 375)

Other solutions are possible.

1. This manual contains removal, disassembly, reassembly, reinstallation, and test procedures.

2. Evaluate the test results in accordance with specification paragraphs 11.9 and 12.2.

3. The support facilities available for this project include a complete CAD/CAM setup, an environmental test laboratory, and a model shop.

4. The sensors feed into a data conversion unit, which will receive, process, encode, and decode the data.

5. The functions of the unit are gathering, processing, and transmitting data.

B-13. *Using Coordination and Subordination*

(p. 375)

Other solutions are possible.

> Direct-mail marketing is an art that few people are aware of. Copywriters know that a reader's eye moves over a page from left to right and from top to bottom. Relying on this conditioned response to the printed page, direct-mail copywriters use graphic features to stop the movement along the path, placing important information at those points. Most readers get the essential message from a well-constructed direct-mail letter without even reading it.

B-14. *Identifying and Eliminating Voice and Mood Shifts*

(p. 376)

Other solutions are possible.

1. Some people think that using the passive voice is always wrong, but those who know when to use it appreciate the contribution it can make to good writing.

2. Rebel against your grammar checker today.

3. Sometimes, writers don't know who performed an action, and they cannot use the active voice.

4. Too many passive constructions makes the reader's job more difficult than it should be.

5. The active-voice sentence is easy to read because it follows the natural sequence of an event.

B-15. Identifying and Eliminating Faulty Comparisons

(p. 376)

Other solutions are possible.

1. An isotope-fueled thermoelectric generator would be more harmful to the environment than a gasoline-powered generator.

2. Fuel cells do not represent an environmental threat any more than does solar power.

3. A full-time gasoline-powered system would be even worse than a diesel-powered generator.

4. Our design poses less threat to the environment than does any other combination.

5. Our design is environmentally friendly, as a good design should be.

B-17. Spelling

(p. 382)

Corrected spellings appear in italics.

accessable	*accessible*	freight	C
accomodation	*accommodation*	further	C
accredit	C	harrass	*harass*
accummulate	*accumulate*		
acetic	C	occured	*occurred*
acknowledgment	C	occurence	*occurrence*
appraise	C		
apprise	C	perogative	*prerogative*
ascent	C	perservere	*persevere*
ascetic	C	personell	*personnel*
assent	C	practicle	*practical*
		precession	C
cite	C	principal	C
comparitively	*comparatively*	principle	C
concientious	*conscientious*	procede	*proceed*
conciet	*conceit*	procedure	C
consul	C	reminiscent	C
counsel	C	remittence	*remittance*
decent	C	schedule	C
deductible	C	sieze	*seize*
dissent	C	site	C
embarass	*embarrass*	stationary	C
enviroment	*environment*	stationery	C
		supersede	C
farther	C	tangible	C
formally	C	technition	*technician*
formerly	C	transition	C
fourth	C		

A Text/Exercise in HTML

Introduction to HTML

HTML, *HyperText Markup Language*, is a *text marking tool* used to prepare information for presentation on the **World Wide Web** (WWW). There is no mystery involved, no need to learn a complex computer programming language. All that is required is the ability to apply a few text ***tags*** or *styles*.

For example, if you want to present the heading and paragraph above on a Web page, you would insert a few HTML conventions (tags, that is) so that the marked text would look like this:

```
<html>

<head>

<title>Introduction to HTML</title>

</head>

<body>

<p>HTML, <u>H</u>yper<u>T</u>ext <u>M</u>arkup <u>L</u>anguage,
is not a language at all. It is, rather, a <i>text marking tool</i> that is
used to prepare information for presentation on the World Wide Web
(WWW). There is no mystery involved, no need to learn a complex com-
puter programming language. All that is required is the ability to apply a few
text <i>tags</i> or <i>styles</i>.

</body>

</html>
```

The characters and words within the angle brackets (<>) are the tags needed to transform the text into a form suitable for Web input. They may seem to be nonsense, but when taken a line at a time, the logic becomes clear. The phrase "text marking tool" is enclosed by <i>, which tells the **browser**[1] that the text following should be displayed in italics, and </i>, which says "stop italics." (The "/" character is used to close out a tag.) Similarly, the other tags represent formatting that we want to see in our final Web page.

Submitting this coded text to the Web, you would see this display:

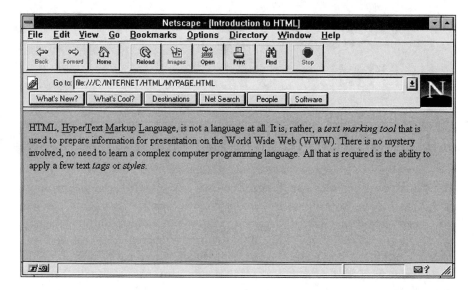

To test the tags and a number of others as you develop them, you will be constructing a simple Web **page**.

In this Appendix, we will cover the defining elements of HTML presentations: hyperlinks and graphics.

In any HTML presentation for the World Wide Web, there are three elements that you should include in every Web page:

- *html*, the document type;
- the *head* and its corresponding *title*, a description that appears at the top of every Web page; and
- the *body*, which encloses the entire content of the presentation.

You should also become familiar with the closing element for most tags, the virgule or forward slash (/). This symbol tells the browser, "OK, stop whatever tag is now in effect." For example, when you mark a word or phrase as

---

[1] A browser is a computer program that recognizes HTML code and translates it into a Web presentation.

boldface, you have to tell the browser when to stop the boldface. If you don't, the browser, not being a mind reader, will continue on merrily displaying all the rest of your page in boldface characters.

HTML

The HTML header tells the browser that this document is coded in the HTML language, rather than one of several other possible document types — the others are not germane to Web presentations.

Head and Title

The Head tag distinguishes the introductory portion of the HTML presentation from the main content of the presentation, the Body. Always enclosed within the Head is the Title tag.

The Head tells the browser that the following information will be the header data for this Web document. The header data is the title that appears at the top of the Web page, usually in the window's title bar. For example, the title in the window displayed above is *Introduction to HTML*. When you select your title, it is a good idea to keep it short enough (fewer than 64 characters) so that the complete title will be shown on the title bar.

The title also becomes a **bookmark** so that when a person performing a search on a topic comes across your page and wishes to return to it at a later time, your title will then be listed in that individual's bookmark list.

Body

The tag Body encloses all the rest of the Web page: text, **hyperlinks**, graphics, animation, tables and forms. The syntax is as follows:

```
<body>

All content except the title

</body>
```

The body can be short or long, plain or fancy, simple or complex. Hundreds of books and **Internet** sites are devoted to Web page design. Unfortunately, it would not be practical to summarize here the profusion of designs that are available.

Commonly Used Text Tags

APPLYING THE HMTL, HEAD, AND TITLE TAGS

Now let us begin to build a Web page using many of the standard tags to embellish the text and insert graphics. We suggest that, if possible, you transcribe the HTML tags on the following pages with a word processor, so that we can demonstrate as we move along.

As we begin, remember the html and Head tags that we have to insert at the beginning of the Web page file.[2] So we start with:

<html>

<head>

<title>We need a title here</title>

</head>

Notice that the Head and Title tags are both closed with the / character. But we won't close the HTML tag until we have completed all our Web page tags. While we're at it, let's change the makeshift title to "My Web Page."

<html>

<head>

<title>My Web Page</title>

</head>

If you'd like, and if you have a browser handy, just combine the HTML lines above in your word processor, then save them as a file with an html **extension**,[3] that is, *yourfilename.html*.[4] So give your file a name, perhaps *mypage.html*. Don't forget that the closing tags </body> and </html> have to be included. Now, open your browser, and open *mypage.html*.

---

[2] HTML tags are not case sensitive. For example, the tag "HTML" is the same as "HtMl or hTMl."

[3] If your word processor cannot save files in HTML, you may want to use an HTML converter. We use HTML Writer by Kris Nosack, a graduate student at Brigham Young University. It's available on the World Wide Web at: http://lal.cs.byu/edu/people/nosack/

[4] Some systems limit file names to eight characters and file extensions to three characters, mainly PCs running DOS or versions of Windows prior to Windows 95. If you are using one of these systems, use a short file name and the file extension .htm.

Setting Up the Body of the HTML Document

Continuing with our Web page development, let us add a few lines of text. Remember, we have to start this part of the page with the tag Body.

<body>

<p>Here is my own Home page!</p>

<p>Don't you think it's great??!!</p>

Here's what the page looks like at this point.

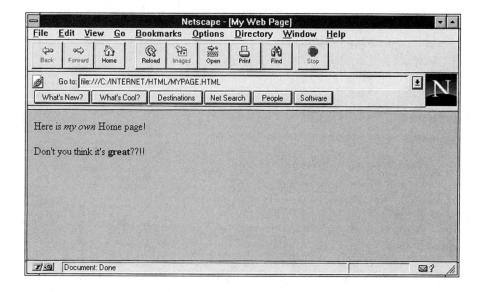

CHARACTER FORMATTING

To set off words or phrases in our text, we need to look at ways to highlight the text. The easiest way to do that would be with the Bold and Italics tags, in much the same manner as in the introductory paragraph. We can modify the line "Here is my own Home page!," emphasizing "my own." Here's how:

<p>Here is <i>my own</i> Home page!</p>As for boldface, why don't we take the next line and make "great" bold:

<p>Don't you think it's <b>great</b>??!!</p>

The gist of character-formatting tags is that you "surround" the selected text with <n> and <n/>, where n is the type font you want.

HEADINGS

Headings used sparingly serve to divide your Web page into logical section, thus:

```
<p>And now for a large headline</p>

<h1>An Eye-Catching Headline</h1>
```

Headlines in HTML are numbered h1 through h6, in decreasing level of importance (and appearance). The <h1> used here is the boldest and largest type available.

Here, once more, is how the HTML script looks at this stage:

```
<html>

<head>

<title>My Web Page</title>

</head>

<body>

<p>Here is <i>my own</i> Home page!</p>

<p>Don't you think it's <b>great</b>??!!</p>

<p>And now for a large headline</p>

<h1>An Eye-Catching Headline</h1>
```

And as displayed on a browser.

Is your home page what you expected? Is there a problem? If so, you should check your HTML tags, line-by-line and character-by-character, because computers expect absolute precision. A missing tag or even a missing < or > will create an interesting — and usually useless — presentation.

Hyperlinks

The **hyperlink** is the single outstanding feature of the Internet. Using hyperlinks, the Internet is able to **link** Web documents anywhere in the world. Or in any folder or directory on your own computer.

Our task is to learn how to construct a few of those links. Again, just a few simple tags are required. First, we will build links within our Web page and then a few links to other Web sites.

To lead into our discussion of links, we will add these two lines to our "script."

```
<p>Now it gets a little tougher.</p>

<h2>Links</h2>
```

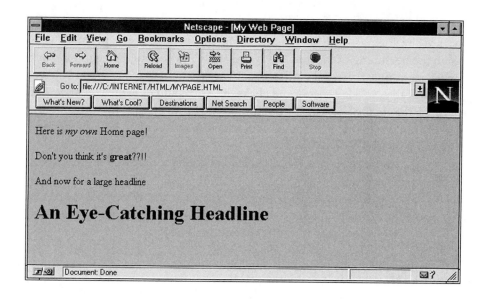

The tag <h2> displays a secondary heading. There are two basic sources for hyperlinks: files and graphics on your own computer, and files and graphics on *any* computer attached to the World Wide Web.

INTERNAL LINKS

Every Web page should have at least one internal link, and that link is a return to the top of your own Web page. The reason for this link is that, when your page takes up several screens (as is often the case), it's easier for viewers to click on the "Home" link than to use the up-arrow or page-up keys to return to the beginning of the page.

Or if you have built other Web pages that are stored on your computer, you can use the Home link in those pages to return to your main (home) page. Here is the way to set up the Home link in our page.

```
<a href="mypage.html"Go home!></a>
```

The <a> tag is an "anchor" that you will find in all hyperlinks. *The href tag is a signal to the browser that what follows is a link to another page, file, or graphic.* Every jump to someplace else is triggered by href.

Notice also that the file name in quotes, "mypage.html", is the very file we are developing — the browser reads the file name and says to itself, "Let's go to the beginning of this file." (A browser always goes to the beginning of a linked page, unless you tell it otherwise.) The "Go home!" request, you will see, is the link that the viewer sees and clicks on. You could use any words you like here — "fat chance" or "dnsa wpodd" and the browser will still return to the top of your page. But "Go home!" will probably mean more to your viewers, who *do* care what you say.

Now, combining all the HTML lines that we have used so far will give us this presentation:

Note the underlined text, <u>Go home</u>!. That is a World Wide Web visual cue that signals a hyperlink. So, generally speaking, when you see underlined text, you can click on that text and jump to another location.

EXTERNAL LINKS

As with the internal links, the syntax is the same — it's just a bit longer. That's because we have to tell the browser where in the world to find what we're looking for, literally. That means a reference to some other computer somewhere and then to a detailed pointer to the appropriate file on that computer. Here is what an external link to the home page of the British House of Commons in London looks like:

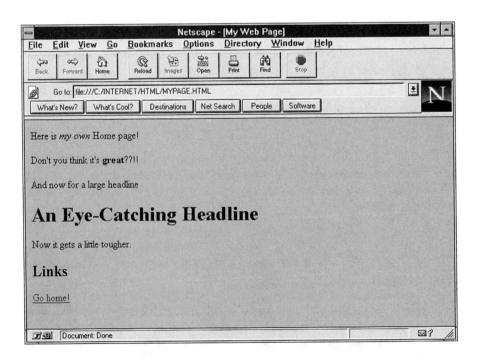

```
<a href="http://www.open.gov.uk/commons/commons.htm"></a>
```

This link, which is known as a **URL** (Uniform **R**esource **L**ocator), will take you directly to the home page of the House of Commons. If we put in an appropriate "jump word" — <u>UK Government</u> ought to do — viewers can then get to the House of Commons page. Thus:

```
<a href="http://www.open.gov.uk/commons/commons.htm">UK
Government</a>
```

Here is what the tags in this link mean:

- *a* and *href* are explained above (see Internal Links).
- *http* is the signal to the browser that says, "Get ready to go to another computer." It stands for HyperText Transfer Protocol and is used in all external links.
- *www* is, of course, the World Wide Web.
- *open.gov.uk* is the name of the faraway computer (in London probably) that the browser will try to find. This part of the link is known as the "domain."
- *commons* is a directory or folder on the London computer.
- *commons.htm* is the HTML file that contains all the data on the House of Commons.

Suppose we place one or two more external links in our Web page that our viewers can visit with a mouse click. (Try them yourself to be sure they work.)

So, where shall we go? Are you interested in golf? It so happens that there are a multitude of links to golf sites (a *site* is the same thing as a Web page).

```
<a href="http://golf.com/links">Let's play golf!</a>
```

Then, maybe you would like to find out something about movie stars and the films they've appeared in.

```
<a href="http://pibweb.it.nwu.edu/~pib/movies.htm">And let's go to the
movies</a>
```

These two sites and the hyperlinks they contain will give you more information on golf and movie stars than you could ever hope to use.

To conclude this section, let's review our HTML file. Here's what it should look like at this point (precisely, by the way):

```
<html>

<head>

<title>My Web Page</title>
```

```
</head>

<body>

<p>Here is <i>my own</i> Home page!</p>

<p>Don't you think it's <b>great</b>??!!</p>

<p>And now for a large headline</p>

<h1>An Eye-Catching Headline</h1>

<p>Now it gets a little tougher.</p>

<h2>Links</h2>

<p><a href="mypage.html">Go home!</a><p>

<a href="http://www.open.gov.uk/commons/commons.htm">UK
Government</a>

<p><a href="http://golf.com/links">Let's play golf!</a></p>

<p><a href="http://pibweb.it.nwu.edu/~pib/movies.htm">And let's go to the
movies</a><p>
```

This is how the file looks when it is displayed:

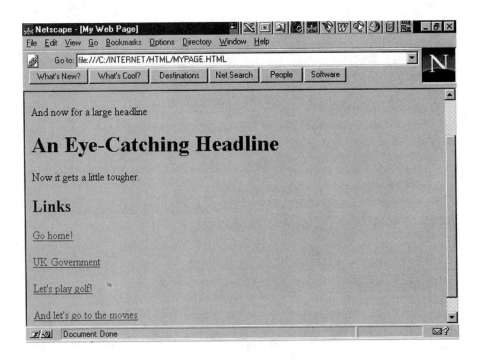

Graphics

Everybody loves pictures. If your Web page *doesn't* have graphics, you won't have many visitors. Therefore, you should become familiar with the fundamentals of graphic presentation on the Web.

If you are artistically inclined, you can create your own graphic images with any of the excellent visual arts programs available commercially. Or you can convert your photograph collection to Web images using a scanner (which is a relatively expensive piece of equipment). You can even buy a relatively inexpensive digital camera to take original photographs that will be Web-ready.

However, most of us have to rely on simpler methods of finding and incorporating graphics into our Web pages. One method that works, even for unartistic people, is to use a graphic consisting of text on a graphic background. We created one such image using a program that came with our word processing package. We then converted the image to Web format and included this instruction in our HTML script:

```
<p>Here's a graphic I put in here!<p>
```

```
<img src="htmlsign.gif">
```

Another handy method is to capture an image on the Web, *one that is not protected by copyright*.[5]

Here's one we found, to which we added an introductory line, like this:

```
<p>A place I'd like to visit on my next vacation<p>
```

And then the call to the image itself with this line:

```
<a><img src="captour.gif"> </a>
```

It appears that we that we now have a new tag, "img src," which looks like, and is, image source. What the tag is saying is to look for a graphic in the "gif" format (which we describe next) and bring it into our presentation.

CONVERTING GRAPHICS TO THE WEB FORMAT

Actually, the photograph we just put into our Web page, captour.gif, was cut out of a Web site with a graphic converter program. Most modern graphics-making programs can convert any graphic format (such as cdr, bmp, drw, tif, cgm, and the like) into one of the formats that the Web recognizes, namely, gif (graphics interchange format), developed for CompuServe presentations.

---

[5] Abuse of copyright laws and regulations on the World Wide Web is, unfortunately, widespread. We urge you to treat others' intellectual properties as you would your own, namely, with respect and in accord with all civilized customs and regulations.

If you are using older programs, you will find many shareware[6] converters available on the World Wide Web. Two common ones are Paint Shop (for Windows) and GifConverter (for Macs).

Generally, a converter allows you to "capture" an image from an outside source, then convert that image to the appropriate gif format. In the process you can modify the image to suit your own taste or purpose.

GRAPHIC LINKS

One more action you need to cover before leaving your Web page development project is how to use a graphic as a hyperlink. If you have accessed Internet Web pages, you know that a good many links are not just underlined text links (as in our Go home! jump). Rather, you are invited to click on a drawing or a photograph to connect to another file or Web site.

The trick here is to change an ordinary graphic into a hyperlink. The procedure is not difficult — all you have to do is *combine* an "href" link with one of your underlined links that has been converted to a graphic format. Since we've just mentioned Go home!, let's use that as a demonstration. We changed the Go home! text into another sign using our simple graphic program, then ran it through our graphic format converter. We set up the href link this way:

```
<p><a href="mypage.html"></p>
```

This tag tells the browser to link to the location that follows, which is our Go Home sign.

```
<img src="go_home.gif"></a>
```

If you'd like to add these links to the HTML script and try it out, you'll find yourself right back at the top of your Web page. In fact, if you design a long Web page that takes up several screens, it's a good idea to insert the Go Home link at strategic places in the presentation so that your viewers can get back to the beginning with one quick click.

Window Dressing

Advanced HTML is nothing more than adding to the tags discussed here. Adding color, moving your text and graphics around on the page, applying sound and animation — it's just a matter of building on what you've learned already.

---

[1] Shareware is a term used to describe programs or routines available on the Internet that you can download and use free of charge for a short trial period. If you decide to keep such a program, you should send the author the usually small payment that is requested.

For example, to give your Web page a color background, you can add a background color from a solid color graphic converted to gif:

```
<body background="bgcolor.gif">
```

To color the background right away, you should put this tag in right after the Body tag at the beginning of your script.

There are many sources of building on your HTML knowledge. Any good bookstore will have dozens of titles on advanced HTML. The World Wide Web is also an outstanding source of information on the very latest in HTML technology. A search using the term "HTML" will lead you to a long list of eminently useful titles.

As for the esthetic side of design, look again to the Web. Find a presentation that really appeals to you. Then, using the View feature of your browser, examine the author's HTML script to see how he or she designed that display. Rework the script to instill your own talent and personality.

Web Authoring Tools

Every day, more authoring programs give the Web page designer a variety of menu options to choose from to achieve a desired effect. Rather than type in your own tags, you would simply select a tag from a menu featuring several dozen HTML styles. A few examples:

Headings. Click on the level of heading you want, and this is what you see:

```
<h1></h1>
```

Or boldface text:

```
<b></b>
```

Or a link:

```
<a href=""></a>
```

All the author has to do is fill in the blanks between the beginning tags or the quotation marks. It's easy, it's quick.

The Final HTML Script

If you need to check your tags, reproduced here is the complete HTML file. Since we have not yet closed out the HTML and Body tags, we have added the closing tags here:

```html
<html>
<head>
<title>My Web Page</title>
</head>
<body>
<body background="bgcolor.gif">
<p>Here is <i>my own</i> Home page!</p>
<p>Don't you think it's <b>great</b>??!!</p>
<p>And now for a large headline</p>
<h1>An Eye-Catching Headline</h1>
<p>Now it gets a little tougher.</p>
<h2>Links</h2>
<p><a href="mypage.html">Go home!</a><p>
<p><a href="http://www.open.gov.uk/commons.htm">UK
        Government</a></p>
<p><a href="http://golf.com/links">Let's play golf!</a></p>
<p><a href="http://pibweb.it.nwu.edu/~pib/movies.htm">And let's go to the
movies</a><p>
<p>Here's a graphic I put in here!</p>
<img src="htmlsign.gif">
<p>A place I'd like to visit on my next vacation</p>
<a><img src="captour.gif"></a>
<p><a href="mypage.html"></p>
<img src="go_home.gif"></a>
</body>
</html>
```

Glossary

Bookmark. A specific location in a World Wide Web page that can be electronically referenced from another Web location, either from within the same file or from a remote Web site.

Browser. A computer program that accepts HTML scripts and converts them to World Wide Web visual presentations.

Extension. The suffix assigned to a computer file that identifies the type of file. In some earlier IBM-compatible machines, the extension is limited to three characters.

HTML. A text and graphics marking scheme in which text, graphics, and links are formatted for processing by a browser.

http. **H**ypertext **T**ransfer **P**rotocol. A code that identifies a computer file as HTML-based and capable of linking to other Internet locations.

Hyperlink. A link (q.v.) that permits instantaneous connection to internal or external Web sites.

Hypertext. A generic term for the electronic linking of one World Wide Web location to another, either within the same file or from a remote site.

Internet. A network of computers that can be electronically connected to one another. The repository for the World Wide Web and other services such as Internet Relay Chat and Usenet (both of which are text-based person-to-person "talk" programs).

Link. An electronic connection between files or between computers that allows users to access a remote location.

Page. The generic term applied to World Wide Web displays. It is not physically limited as is a book page, but can consist of a series of display screens.

Home page. The first or top screen in a World Wide Web display.

Web page. The same as "page."

Web site. A complete World Wide Web display (which may consist of several pages or several linked files) located on one computer identified by a unique electronic address (see **URL**).

Tag. A text and graphic formatting characteristic. In HTML, a marking scheme that specifies display characteristics for World Wide Web pages.

URL. **U**niform **R**esource **L**ocator. An electronic address for World Wide Web sites. It consists of the domain name for a computer, directories or folders within that computer, and specific files in the named directory.

World Wide Web. A subsystem of the Internet that permits visual displays and the interconnection of these displays on linked computer systems.

Index

Semicolons, rules for using, 350
Sentences
    complex, 341
    compound, 341-342
    compound-complex, 342
    dynamics of, 342
    simple, 341
    the elements of, 338
    types of, 341-342
Service change bulletin, 5
Service change, 10
Simile, 61
Single meaning, 4
Software user's manual, sample
    pages, 246-249
Spatial mode of exposition, the, 67
Specification, 49, 110, 111
Speeches
    extemporaneous, 280, 282
    impromptu, 280, 281-282
    industrial, 280, 283
    read, 280, 282
    structure, 284-286
    visuals for, 284
Spell checkers, computer, 25, 26
Spelling, 89, 120, 377-382
Split infinitives, 99, 100
Standard English, 15, 17, 18, 19
Straw man, 74
Style guides, 110, 116
Subordination, 97
Suffixes, 378-379
Syllogism, 70
Symbols, 117, 118

T
Tables of contents, 180, 181
    in proposals, 209
    sample of, 189
Tables, 87, 88, 130, 131
    planning and coordinating, 82
Technical editor, 34, 41, 33
Technical expert, 42

Technical illustrator, 34, 42
Technical terms, 67
Technical writing group, 31
Technical writing team, 31
Test procedure, 9
Title page
    in proposals, 209
    sample of 187
Tone, 3
Top-down development, 80
Trivium, the, 84

V
Verb, definition of, 333
Verbs
    characteristics of, 335-338
    choosing powerful, 86
    mood, 336-337
    number, 335
    person, 335
    tense, 335-336
    voice, 336
Voice shifts, 96, 97, 371-372

W
Warnings, cautions, and notes, 239,
    240
*Webster's New International
    Dictionary, Second Edition*, 18
*Webster's New International
    Dictionary, Third Edition*, 18
Weekly production report, 10
Word processor, 25, 26
World Wide Web, 28, 29

X
Xerography, 25
Xerox 3006 Plain Paper
    Fax/Printer, 23, 24
Xerox 5390 Photocopier, 27, 28
Xerox XPrint 4915 Color Laser
    Printer, 26, 27